FORTY SUNS

The Story Of The New Space Age

David G McDaniel
&
Michael C Petry

Zero Edition

FortySuns.com

"Turning the eyes of this world toward the next."

Copyright David G McDaniel & Michael C Petry

NOTE

This may be the fastest, and funnest, book on space you've ever read.

Enjoy.

Copyright David G McDaniel & Michael C Petry

Our ultimate goal with *Forty Suns?*

That you:

1. Read it (please).
2. Enjoy it. Maybe even get inspired.
3. Find yourself, by the end, excited about our space future.

And that perhaps, in reading it, you become a bit more hopeful for both yourself and your fellow humans.

Space is the answer.

Forty Suns: Zero Edition

Copyright © 2022
David G McDaniel & Michael C Petry
Reprinted, Copyright © 2024 R2

Published by
Forty Suns

Cover design by
Forty Suns

All rights reserved. This book may not be reproduced
in any form, in whole or in part, without
written permission from the authors.

Visit:

FortySuns.com

"Transformative."

"As much a book about our human capabilities as it is about space. *Forty Suns* will shift your view as to what's possible."

"There may be hope for us yet."

Forty Suns: The Story Of The New Space Age

FOREWORD	14
A Worthy Note	21
CHAPTER I	23
The Next Frontier	23
Where We're Headed	23
The Challenge	25
Making The Dream Real	26
A World Of Fans	27
Opening The Chamber	28
Why "Forty Suns"?	30
Imagine This …	32
CHAPTER II	34
Our Space Future	34
The Concept	34
STEM	36
Roman Locomotives	39
Eurekas & Creative Bursts	40
The Reality Facing Us	42
Nothing Is Making Us	44
Ad Astra	45
Checking System Pressure	47
T-minus 3 … 2 …	48
CHAPTER III	51
Power	51
Viva La Revolution!	52
Outputs	53
Torque	54
Electrical	54
Thrust	56
Sources	59
Batteries and Storage	59
Combustible Fuels	62
Nuclear Fission	64
Nuclear Fusion	65
Ready-Made	71
Power As A Yardstick	73
The Illusion Of Scarcity	74
CHAPTER IV	76
Space	76
Forces	77
Survival	78
Distances and Travel	79
Moving In 3D	80
Space Age 2.0	82
What Are Some Realistic Objectives?	82
Lunar & The Artemis Program	83
Orders Of Magnitude	85
Estimation Of Effort	88
Corporate Influence	88

Copyright David G McDaniel & Michael C Petry

Space Mining	89
Commercial Space	91
To Infinity And Beyond	92
Space Is Hard	94
CHAPTER V	**97**
Habitat & Transport	97
Orbital Operations	98
Sustainability	101
Living & Working	103
Getting Around	104
Zero Gee, No Surface	105
Low Gee, Surface & Air	106
SOPs For An Extraplanetary Existence	108
CHAPTER VI	**110**
Terraforming	110
Mars	110
Venus	111
Closer To Home	113
One Day	114
CHAPTER VII	**116**
Planets & Cosmology	116
Planet Hunters	118
Life As We Know It	120
Our Own Backyard	120
Abundance	123
Pure Discovery	124
CHAPTER VIII	**127**
Physics	127
Universal Constants	129
The Big & The Small	130
Fancy Tools	132
Fiction Becomes Reality	134
CHAPTER IX	**137**
Chemistry & Materials Science	137
Strength	137
3D Printing	138
Game Changers	139
CHAPTER X	**143**
Robotics & AI	143
Our Mechanical Companions	144
Power Up	145
The 3 Laws	147
The Digital Mind	148
Types Of Artificial Intelligence	149
Touched By AI	151
AI & Space	153
Not The End, The Beginning	155
CHAPTER XI	**157**
The Human Condition	157
The Outer Limits	158
A Few For The Books	159

Copyright David G McDaniel & Michael C Petry

Ghosts In The Shell	164
Bionics	165
Cybernetics & Cyborgs	166
Enhanced Organics	167
Radiation	170
Our True Measure?	172
Us Among The Stars	174
CHAPTER XII	**177**
Communications & IT	177
Home Grown	177
Orbital	180
Latency	181
Networking	183
More Power!	184
Our Quantum Future	186
Efficiency vs Demand	187
The Big Picture	188
Alternate Reality: Eyes On The Prize	189
CHAPTER XIII	**192**
Education	192
Seeing To Yourself	192
Encouraging Others	193
Supporting STEM	195
CHAPTER XIV	**198**
Governance	198
Treaties & Accords	199
The Artemis Accords	200
Keeping Pace	201
Bureaus, Councils & Departments	203
State Players & More	204
Eternal Vigilance	206
The Best Of The Best	209
Make Your Voice Heard	210
CHAPTER XV	**212**
Progress Is Possible	212
A Sprint, Not A Marathon	213
The Tipping Point	215
War! Good God Y'all!	216
The One Tenet	219
The Bigger Picture	221
United We Stand	222
Getting There	223
New Milestones	224
Can-Do	227
Audacity	232
Naming A North Star	234
Character	235
Doing Our Part	236
CHAPTER XVI	**237**
Turning Point: Space	237
The Upward Spiral	237

Thinking Extraplanetary	238
Mindset	240
The Money In Space	241
An Example	242
Demand	245
Space Vacations?	247
Space: The Great Unifier	250
Group Power	251
Science Fiction	253
A Global Effect	253
Space Opera vs Fantasy	254
By the Numbers	256
CHAPTER XVII	258
T-Minus Zero	258
Level Up	260
Space Lord	261
Space Captain	261
Space Marine	261
Space Fan	262
Space Newb	263
Formula For Success	264
Our People	264
Get 'Er Done!	265
A Few Thought Exercises	266
The World Is Your Oyster	266
Too Much Help?	266
Life Outside	267
Despicable Me	269
Expanding The Wins	270
One Last Sports Analogy	270
The Struggle Is Real	271
How?	272
Adieu Ado	274
The Bullets	276
Has Anything Shifted?	277
Destination Space	278
About Your Authors	279
APPENDIX A	280
What Now?	280
The Cosmos	281
General Science	282
Rockets & Launches	282
Space Agencies & Programs	283
Space Commerce	283
Ways To Get Involved	284
Ways To Become Informed	285
Ways To Help	286
APPENDIX B	290
Foundation	290
Forty Suns	291
Thee Analogy	291

Copyright David G McDaniel & Michael C Petry

- Who Dares... 292
- APPENDIX C... 294
- What Might The Future Take?... 294
  - The Moonshot Mindset... 294
  - Responsibility... 295
  - Ethics... 297
    - Greatest Good... 298
  - The Elephant In The Room... 299
    - CAVE People... 300
- APPENDIX D... 302
- Brave New Worlds... 302
  - The Science Of Failure... 303
  - Reality Check... 305
  - Adjusting Our POV... 306
    - Through The Eyes Of Others... 307
  - Beating The Apocalypse... 308
    - The Lazy & The Afraid... 310
    - Gamification... 311
  - The Brave & The Bold... 312
- APPENDIX E... 313
- The Gettysburg Addendum... 313
  - Risk vs Reward... 314
  - Optimism... 316
    - The Stakes... 317
  - The News... 318
    - It Ain't As Grim As All That... 320
    - For Your Consideration... 321
    - Morale... 323
    - IIS... 324
    - Conflict Sells... 325
    - Human Nature... 327
    - What If ... ?... 329
    - Our Native State... 330
    - Culture... 332
  - Perspective... 334
    - Our Brotherhood... 335
    - Taking Ownership... 337
    - The 90%... 337
  - On What Could We Agree?... 339
    - A Hypothetical Round Table... 340
    - The Value Of Talking... 340
  - Freedoms... 340
  - Soul Of The Frontier... 342
    - Together... 343
    - Why We're Here... 345
  - Code Of A Future-Builder... 346
  - From Us To You... 347

Dedicated to the continuum.

This is our playing field.

Have fun while you're here.

Leave it better than you found it.

Don't get too serious.

"Hello wonderful person …"

Anton Petrov

FOREWORD

THIS IS A BOOK ABOUT SPACE.

It's also a book about ways we humans can work together.

In most cases the pursuit of space, along with increasing our presence in space, is a thing on which many of us could agree.

A discussion of space therefore:

1. Serves to raise awareness of the very real need to expand into this next frontier.
2. Provides common ground over which to connect.

Forty Suns is a book about space.

Within that discussion we may learn a bit about ourselves.

—

Space has captivated us since the dawn of time.

Our ancestors may not have known it as "space", but the fascination was the same. For them it was the home of the gods, brilliant points of light in the night sky, dynamic, celestial entities, moving slowly over time, planets, constellations, bands of color imbued with all sorts of imaginative explanations. It was a place where fantastic events were happening, high above.

Whatever our ancestors thought, one thing was certain:

That thing, space, represented an untouchable vastness.

It also represented a thing common to each of them.

Later, as we came to understand more of the world around us, we began wondering if it might be possible to go there. To experience those faraway places. Not just look at them through a telescope, but actually travel out there, among the stars. To go to space. Musings which have accelerated faster toward reality in the last century. Humanity has made greater strides toward that objective in the last short span of decades than in the entire rest of our existence.

Now we've done it. We've gone to space.

Our next generations will live and work there.

So far we've put handfuls of us on the nearest celestial body, the moon, and are looking seriously at going further. It's important that we do. Because Earth, our wonderful home—we now realize—is but one tiny dot in that sea of vastness.

Not quite as eternal as we once might have believed.

Until very recently, the best we could hope for was to take care of it, while simultaneously hoping it didn't get smashed by a big rock.

That wouldn't be great for any of us.

Worse, big rocks are just one way our cozy cottage in the cosmos could get ruined. No matter how well we see to it, our wonderful world could burp, or shrug, and kill us all. Not that the Earth is unkind, it isn't, only ... it really doesn't care.

Narrating the documentary *Life on Our Planet*, Morgan Freeman names one of the Rules of Life:

"Earth never remains stable for long. Sometimes that helps life, sometimes that hinders it."

Ancient-us couldn't do much about extraplanetary options.

Modern-us can.

Across the long history of life here on Earth there have been five major extinction events, each driven by entirely natural (and in each case different) factors.

1. 440 Million Years Ago (MYA), 86% of life on Earth lost.
2. 365 MYA, 75% of life lost.
3. 252 MYA, 96% of life lost. This one was the worst.
4. 201 MYA, 80% of life lost.
5. 66 MYA, 60-75% of life lost. That was the dinosaurs.

And we may be in the middle of the next one right now. To put it mildly, stuff happens. Whether we have any say in it or not.

Meaning we need options.

Soviet rocket scientist, Konstantin Tsiolkovsky, phrased the situation nicely back around the turn of the last century:

"Earth may be the cradle of humanity," he said, "but we can't stay in the cradle forever."

Much as we love it, we've needed alternate solutions to living on Earth since ... well, since we've been here. Since long before Tsiolkovsky made that astute observation. A "second home" option, if you will. Getting out of the cradle, however, has never been a thing we could realistically even dream of.

Until now.

Thankfully we made it this far. No burps. No civilization-ending shrugs. No big rocks. No other things we didn't manage to deal with and keep advancing. Through the millennia of turbulent expansion, through all the rough spots of existing as humans on Earth, we've arrived. Here, now, where we at last have reason for *real* hope. Finally, with a concerted effort, we can solve the "one world" dilemma that's loomed over our heads since the beginning.

Finally we can conquer space.

While taking better and better care of this wonderful planet we call home we can, at the same time, realistically set our sights on expanding our possibilities. The technology to do so is being developed all around us, faster than at any point yet. Space, and living elsewhere, is, for the first time in our history, possible.

A reality many may not be aware of.

Forty Suns exists to fill that awareness gap. It is simultaneously a short and a long look ahead. The journey of a thousand steps begins with one, as a wise man once said, and if our journey toward an extraplanetary existence will be a thousand steps, the steps we take today are the ones that will eventually get us there.

But those steps must be taken.

Good news is, they are. Each and every day, by pioneers in the field, by newly formed and forming enterprises, by national interests, scientists and explorers and on and on.

If you stop and look, you'll see space is being worked toward by *way* more people than you probably realized. In fact NASA recently heralded this as the next generation of space exploration.

We are, whether you've observed it or not (and *Forty Suns* will fix that if you haven't), at the dawn of a new Space Age.

Call it Space Age 2.0.

People already have.

Part of this has to do with opportunity. In particular economic opportunity. A lot of players (and there are lots) are currently focused on the space economy. Those numbers have been increasing exponentially in the last decades, which is good, because money drives progress.

Our dreams may be noble, but our reality takes cash.

Cash (profit) is the battery that runs the world. This stable of new and established enterprises are launching and will launch fresh ventures, while at the same time advancing the world's technologies in their quest for viability.

That growing economic interest, combined with an explosion of new programs being put into place by nation states around the world—plus a rising groundswell of global interest from space enthusiasts like we and thee—is leading, even now, to a renaissance of our space future.

These realities are fantastic.

They need to be strengthened.

Let's put it another way.

If our focus determines our reality, it stands to reason that by focusing on space it will ...

Become our reality.

It's not enough that we simply survive. At least not on a global scale. Individually yes, perhaps. Most of us (your authors included) work toward our mortgage payments, hoping to save enough for a better TV. Maybe a vacation.

We survive.

And that's fine for most of us. As a planet, however, the only way we achieve the next frontier is to survive *in abundance*. Across the board our thinking, our planning, our reach serves us best when we direct it toward the seemingly impossible.

Go big or go home.

A popular saying.

Maybe we should change it to "Go big or go bigger."

Why give ourselves the other option?

Striving for an abundance is the way we get there.

Forty Suns means to be the focus for that push. A consolidated overview of the things that will get—and are getting—us there.

Space.

The short and the long view. Things our fellow humans around the world are doing, other things not yet being done but that will be done, advancements to pave the way. Even things that are only just now being imagined.

Especially that.

For in order for our space future to become real, the ability to imagine things hitherto unimagined, the determination to believe in—and reach for—things once believed impossible, will be key.

The impossible will set us free.

It's been said every breakthrough was a crazy idea before it became a breakthrough. Incremental improvements are good, even necessary, but we need a healthy dose of crazy ideas.

Audacious ones.

A quote from one of the Wright brothers, first humans to fly (controlled flight, anyway; no doubt a few of us have flown via catapult or other means down through the ages), gives a nice encapsulation of that driving premise:

"If we all worked on the assumption that what is accepted as true is really true, there would be little hope of advance."

In other words, to go big (or bigger) we have to think big.

Our main goal, then, is to start you, and the world, thinking. To do that we cover a lot. Herein we discuss the entire field of space exploration, commercialization and colonization, along with how we humans are making, and will make, that happen.

You'll note it's not that long of a book.

Which means the discussion is very high-level. Our sole purpose is to inspire. If successful *Forty Suns* will act as a springboard, sending you eagerly off to discover new interests, to do your own research, make your own space connections and, ultimately, learn more about—and possibly even support—humanity's prospects on this final frontier.

Be prepared, then, for a fast, sweeping run through the entirety of Space Age 2.0. When done you'll have a clear overview of everything that's happening.

Where you go from there will be up to you.

Of course we'll wrap things at the end with a bunch of ideas for that, too. Going deeper along the way on anything you read is just a phone or computer search away.

Our hope for this book is that it be:

- Enlightening
- Inspiring
- Fun

And maybe, just maybe, that it gets you pondering what your role in our space future might be.

What can *you* do?

How can *you* be a part?

Hm.

Forty Suns, therefore, is founded on the idea that, at times, one may do more for the advancement of positive change by posing a question than by answering one.

—

To rephrase it, the book in your hands won't teach you all there is to know about space, but it *will* arm you with curiosity. And that may be the most powerful weapon there is.

Who knows?

A nugget or two within these pages may be the inspiration that launches the career of a few new space cadets.

A heroic *Forty Suns* victory cry might go something like:

"From Space Newb to Space Nerd in a single read!"

Or, for those with a preference:

"From Space Greenhorn to Space Geek in a single week!"

Could one of those new players be you?

After all, the best way to predict the future is to create it.

Forty Suns is the starting point. The launch pad, to use a more fitting term. The humble place from which you embark on your journey to become one who knows.

One who knows what?

- Where we're at in the new space age.
- Where we're going.
- How space is changing the world for the better.
- How to keep up.
- Who the major players are now.
- Who they will be in the future.
- What it will take to become extraplanetary.
- What you can do.
- Why each of us matter.
- Why space is important.

We won't answer every question we stir up. See the above. No, if we do our job you'll be putting this book aside, impatiently reaching for all sorts of *real* resources on your favorite space topic.

So much is being done.

Advanced, way advanced things, way incredible projects, missions, exploration, science and discoveries. More good things are happening right now in the pursuit of space than you quite likely realize. More progress is being made, on more fronts, than most people know.

Right now, even as you read this, there are amazing humans out there doing amazing things to advance the cause.

To help us survive in abundance.

So let's turn our eyes toward discovery. Awareness. Let's explore the realities and opportunities of what might be our collective space future.

It's a grand one. And it's already happening all around us.

Welcome.

Your Authors

A Worthy Note

WORDS, WORDS, WORDS.

Before you get to reading, a brief note.

As mentioned this book isn't particularly long, as books like this go, but it's still got a crap ton of words in it.

Most of them you'll know. Some you may not.

In most cases we've tried to define words as we use them, when we think they may not be that common, but if you see one you aren't sure of, take a quick sec to pull out your phone and check it. We're not overly science-y here, but this *is* a book on space, so not all the words are normal everyday stuff.

Example, Plasma. We use that one later in the section on fusion power, but in looking things over it doesn't appear we took the time to define it.

Bad us.

But there's still hope. If you look up "plasma" online you find:

"Plasma is one of the four common states of matter—solid, liquid, gas, and plasma. Plasma is an electrically charged gas. Because plasma particles have an electrical charge, they are affected by electrical and magnetic fields. This is the main difference between a gas and a plasma."

See? That kind of info might help with understanding.

In fact, let us define a word we used right here, up above in this Worthy Note.

Crap-ton.

It means "a lot".

Happy reading.

The future is now.

CHAPTER I

The Next Frontier

IT'S A GREAT TIME TO BE ALIVE.

Some will take issue with that statement. Some always do. We stand by it. Sure, we may have kicked things off in the Foreword with a grim discussion of mass extinctions, but the fact is the promise of being able to extend our reach beyond these shores has never been greater.

Welcome to *Forty Suns*.

A concise overview of space. How much does the average person know? So much is happening in the field of space, much of it unfolding unmentioned behind the headlines ... putting together such an overview seems timely. *Forty Suns* is here to present the projects and technologies driving the pursuit of space in a way that's accessible to everyone.

Ready?

Where We're Headed

IN THIS BOOK we take a look at what's happened, what's happening, what's coming and what *could* be coming. We even do a little peering ahead to imagine where we might set our far-distant sights. In that sense you could say *Forty Suns* is a "compilation of the foundations of our collective tomorrow".

Fancy?

We think so.

The objective?

To unite the focus of humanity toward common extraplanetary goals that create an ever-improving future, by building upon the things we hold in common despite the things we hold different.

This is based on the idea that the average person has little to no awareness of the latest in space technology, engineering and science—and yet would be *very* interested to know. *Forty Suns* aims to raise that awareness, along with overall interest in a broad range of related areas.

There's a Part Two to that mission:

To maintain focus on these pursuits, to ensure they do not slip, that they are not forgotten, that they remain at the forefront of our collective minds.

Because, well, sometimes we forget what we were doing, and things get put to the side.

Back in the early 70s, when your authors were wee lads, the first Space Age was in full swing and we eagerly looked to an amazing future, with flying cars and bases on the moon. We drank our Tang, had plastic rockets (plus actual working ones—remember Estes model launchers? Frickin fun to watch those suckers go off), every bit of NASA paraphernalia and a ton of hopeful imagination. We *knew* our future was in space.

Of course that didn't quite pan out.

All we realized at that tender age was that the dream had simply faded. The whole thing, at least the epic things we were expecting, more or less fizzled.

Today there's new hope. Though we're now a lot older (a crap-ton older?), a new space age has been launched, and there's once again hope for seeing those far-out things in our lives. Example: legitimate flying cars are more or less here. Called eVTOLs in their current iteration (electric vertical take-off and landing), they'll soon be running test flights in urban areas. The old refrain, "We were supposed to have flying cars by now" was such a common complaint for so long, uttered by those (like your authors) bemoaning the fact that our space future never happened, we might take "flying cars" as a subconscious marker of having made it. By that marker, then, with legitimate flying cars just around the corner, we feel confident declaring the next—and this time, we hope, here to stay—space age is upon us.

Let's not let it slip.

Now that Space Age 2.0 is here, let's keep the coals on the fire.

If for no other reason, do it for us.

The Challenge

MASTERING SPACE WON'T BE AN OVERNIGHT SUCCESS.

Those of us who go won't trade walks in the park for more walks in the park.

Space stations are cramped.

Rocket capsules even more so.

Living on the moon will bring with it all sorts of discomforts and challenges, sacrificed conveniences and so on.

Mars will be even worse. The first few generations of settlers on Mars will endure all kinds of hardships, risks—every minute of every day will be dedicated to staying alive.

Space is unforgiving.

Space is hard.

But we're doing it.

Future generations will thank us.

Our science-fiction utopias will one day become a reality for those far-distant descendants, thanks to what we do now.

And we owe it to them.

We live in cities and fly between continents because *our* ancestors, at least some of them, made huge sacrifices, pushed boundaries, challenged conventional wisdom, ignored artificial limitations, took risks and dreamed big.

Thanks to them, we live in the civilization we do today.

Thanks to us, our ancestors will live among the stars.

We don't all have to make those sacrifices. Most of us won't.

But we don't have to. What we *do* have to do, or what we *should* do, is throw our full support behind those that will.

Those that are.

In that way our space future will be assured.

Forty Suns exists to rally that support.

We're here to shine a light on hope.

Making The Dream Real

DO YOU LIKE COMICS? Sci-fi?

So many of us these days do.

More and more, if we keep the throttles open, our fictional things will become reality. Right now more is being done to advance our presence in space than ever before, along with all the cool tech that must, by necessity, go with it.

As with the first space age, new technologies will evolve from the push to build our orbital and lunar infrastructure. Pushes to do new things always lead to better/easier ways to do them.

Which means a better world for all of us.

Want a short list?

From the pursuit of space so far we have, among other things:

- wireless headsets
- LED lighting
- portable cordless vacuums
- freeze-dried foods
- memory foam
- scratch-resistant eyeglass lenses
- medical imaging techniques

- artificial limbs
- water filtration systems
- solar panels
- firefighting equipment
- shock absorbers
- air purifiers
- home insulation
- infrared thermometers
- shoe insoles
- invisible braces for teeth
- cordless tools
- satellite navigation

Are we getting closer to science fiction becoming our reality?

Truth is science fiction *always* becomes our reality. That's just the way it goes. Leonardo Davinci would consider modern helicopters pretty science-fictiony, for example.

It's all relative.

Question is always, how fast do we progress? At what point do we move so fast our newest stuff *seems* like science fiction?

Events of the last years may be a harbinger of that accelerating pace. Invisibility tech? It's already being worked on, right now. Tractor beams? That too. Warp drive?

Yep. Ideas are being evaluated and refined.

From here on out we're each likely, at some point or another, to be amazed with stuff being invented or getting done.

Let's keep it rolling.

This is the dawn of our space future.

And our sci-fi one as well.

A World Of Fans

WHEN THERE'S INTEREST IN A THING, that thing tends to gain steam. The NFL is huge because lots of people love watching football. What if lots of people loved following what was happening in the world of space?

To help achieve that *Forty Suns* aims to energize interest, creating a world of space fans. Like sports fans, these would be people who pay attention to and drive demand for their chosen space interest. It might be rocket launches, it might be lunar exploration, deep space probes, satellite arrays, scientific discovery or anything else that's getting us there.

Fans come in all shapes and sizes.

This book, therefore, isn't geared toward scientists and academics—they already love science and space—but is meant, rather, for the rest of us.

In a sense we're "reaching beyond the choir."

Forty Suns should appeal to just about anyone, in particular those who already take even a vague interest in scientific frontiers, discovery, and the expansion of human capabilities. As well, it should be accessible enough to create curiosity where there was none.

Until now most discussions of the topics of space have been too clinical, as if it were a rarified domain. Space *is* a rarified domain, but we believe at this stage in our history nearly everyone will have *some* sort of interest in knowing more.

Everyone *should* have an interest.

And so this is a summary meant for everyone, of where our space future is headed, what it could be, and how we might get there.

Opening The Chamber

SPACE, LIKE ANYTHING, can become its own echo chamber, with its own, isolated groups of enthusiasts.

Those who get it can't understand why those who don't … don't.

We're out to fix that. We're out to see that *everyone* gets it.

A few points.

- Whether you've noticed yet or not, you're already part of a spacefaring civilization.
- So far the "spacefaring" part is small.
- It will continue to grow.
- Now is as good a time as any to:
  - Pay closer attention.
  - Find ways to support that growth.

Space is the answer.

The challenge, of course, is coming together over this one point we all have in common.

Extrapolate natural human frictions across billions of us and it should be no surprise we have trouble getting along. At times it's a wonder we have rockets at all, let alone any of the other amazing conveniences we enjoy as part of our modern civilization. Throw in the longterm evolution of cultures, beliefs, politics and all else, and it's clear we've been busy making this single planet a fun (is that the right word?) and interesting place to compete.

Quite a few complicated games we have going on.

Some are creative and advance our survival, others not so much. By the time you're through this book you will (hopefully) have the idea that maybe, just maybe, we could find ways to work together, as a world, long enough to add extra places to play. Expand the sandbox, so to speak.

If a billion of us truly paid attention—with interest—to what's already going on in the pursuit of space (and there's *a lot* happening), a million of us might get active and get involved.

That would be a major difference-maker.

With such attention and such involvement, our extraplanetary future would quickly become real.

Let's be honest. In the same way it was inevitable humans would end up all over the planet, it's inevitable we'll end up in the solar system and beyond.

That being the case, why be slow about it?

Why "Forty Suns"?

WHY THAT NAME?

This would probably be a good place to clear that up.

The truth is, it came relatively randomly. During a coffee binge, as it will be recalled, trying to dream up a way to frame a common goal for as many of us as possible.

The name does, however, have significance.

Forty Suns is a nod to the fact that our own sun is but one of many, and around those many other suns are worlds like ours. Which is to say there's a much larger playing field out there, and it should be our objective to expand into it. Forty is an arbitrary number, but it works for our purposes. If we knew of or had access to forty new worlds, what opportunities would that create?

Forty Suns exists to remind us that we're larger than our differences. Perhaps more importantly, *Forty Suns* exists to bring us together over common ground.

Space.

See our Mission above.

Any successful group, whether stellar, global, local or otherwise, is made up of agreements. We're all individuals, that's what makes us great, but we're also in this together. That togetherness, the mortar of our "house of humans", relies on seeing eye to eye where we can. When we share common goals and purposes ... well, that's when the magic happens.

Philanthropist, scientist, entrepreneur—basically, one of those superhumans driving progress on this planet—Yuri Milner says this on his site:

"Any organization that is serious about doing something significant has a mission. But human civilization, as a whole, has nothing resembling a shared mission."

A unique and perfect analogy.
He goes on to add:

"In the long term, that means we cannot thrive—or probably even survive. But what could such a mission be? People, nations, and cultures vary wildly. Where on Earth do we look for a common goal?"

We find the answer in the one thing common to all of us:
Space.
Space is mostly virgin territory; a true "new frontier", offering us an opportunity to come together to find ways to solve it.
Occupying 40 suns would certainly be a huge, huge leap for humanity in our current state but, were we ever to achieve that, it would take a whole lot of working together along the way to get there.
So to the question "Why *Forty Suns?*" the answer is:

01. We could use a shared goal like this, and
02. It's a fitting name.

To be clear, *Forty Suns* is based on real possibilities. What's happening now, what could be happening soon. It isn't about *Star Trek*-level stuff in our far future, or even fast interstellar travel in this generation—though we'll talk about such possibilities. It's about spinning the flywheel of current advancement, doing the things now that will get us to the next level and the next so that, one day, that eventual interstellar civilization becomes reality.
The journey of a thousand steps?
This is where it begins.

—

Reaching another sun is a long road indeed. Especially from this starting point. Space is hard. "Doing" space is not for the faint of heart. So much, however, can (and needs to) be done to reach that first next-sun, it's high time we got started.

Space is a goal we can, and should, focus on together.

What can we do?

More specifically, how does one "Forty Suns"?

The answer is pretty easy, and could be pretty fun:

Add space to the mix.

Live your life as you do, do your favorite things, watch your favorite shows, hang out with your favorite friends, go to your favorite restaurants and eat your favorite foods. Among that, throw a little attention now and again toward space things.

You'll be surprised how that builds.

Imagine This ...

HOW ABOUT A QUICK ORIENTATION?

If you're still reading we assume your silent answer to that question is something along the lines of, "Sure, why not?"

Here we go.

Imagine a beach.

Now imagine being there.

Scoop up a handful of sand. Let it sift through your fingers, paying attention to the streams. Now pay attention to the grains. Stop a few; isolate a single, tiny grain and look closely.

Imagine that's a star.

Now look again at the sand you have in your hand. Any remaining, even those little grains stuck here and there. Those, too, are stars. Now expand your vision, back out from that focused view, and take a look at the sand all around. The sand you're standing in, the sand in your immediate vicinity—all the sand you can see.

Stars.

Each grain. All the way down the beach, as far as your vision goes. Each and every grain of sand a star.

Now go further. Imagine every beach in all the world, every desert, every back yard, every grain of sand.

That's roughly how many stars there are in the observable universe. Many of them of a type that could host one or more habitable planets.

Now look again at that single tiny grain you first chose.

That's us.

All the rest of it is what awaits.

Which is to say, there are a near infinity of goals out there to achieve.

The question then becomes:

Who, and what, is getting us there?

CHAPTER II

Our Space Future

DID WE SAY SPACE IS HARD? Look around at our everyday world and consider all the moving pieces, all the things it takes to create and hold together a modern civilization.

Space is the next evolution of that Earth-based existence; a much, much trickier proposition. Conquering it will take moxy.

There was a movie in the early '80s called *The Right Stuff*. It presented a small slice of what it took to make the first national space program work, showcasing the things the people involved endured during an era of great political turmoil and technological innovation.

Sound familiar?

A successful Space Age 2.0 will require us to revisit that sort of collective human determination. That we once again have "the right stuff". Probably focused through the lens of a select few, the brave souls that pioneer the next frontiers; brand new heroes that will come from many fields.

Our space future depends on increasing our knowledge and understanding of the things that will get us there.

The Concept

HAVE YOU EVER SAID TO YOURSELF, "I wish I was more interested in space."?

You may be in the right place.

Allow us to use another of what will soon become our famous (infamous?) sports analogies (you'll see).

Let's say you were interested in becoming interested in football. Not that you were interested already, but that you were interested in *becoming* interested.

Tricky concept, that.

You feel you want to be more a part of the world of football, or at least be more aware of it, yet you can't seem to get yourself interested in getting interested. You see people around who know a few of the teams, understand a bit of the game, check the occasional score and so on. That could be you, you think. There are *plenty* of resources to help you learn about football.

If only you were interested.

And so it is with space.

There are *many* ways to learn all sorts of things about space. That's not the issue. From how rockets work to how we'll live on Mars, information exists in abundance covering thousands of space topics, much of it made fun and engaging for the rest of us to easily understand. It's no longer hard to know.

Remember that old saying? "It's not rocket science!" That was, of course, meant to imply rocket science is hard. And it is.

Only, these days anyone can learn the basics

We're in an age of incredible teaching technology, coupled with unprecedented information availability—along with an abundance of smart people who love to share. You're alive at a time when cutting-edge how-to videos and guides exist, readily accessible (All Hail the internet), with slick production values, clear graphics, fun examples and exceedingly plain-language explanations. You no longer have to be a rocket scientist to understand how rockets work. Or how anything works. There are tons of resources out there that will lead you, often in a quite entertaining way, to the "Ah ha!" moment about the thing you want to know more about.

But you do have to be interested enough to look.

And that's the real issue.

How do you get yourself interested enough to actually want to sit down and learn that stuff? How eager are you to be in the know? The resources are there, free and easily available, that can and will lead you to utter brilliance on the subject of space.

So ...

How do you get interested in being interested?

That's where we come in.

We might summarize the stages of interest with a scale:

1. Not yet interested.
2. Interested in knowing more.
3. Actively learning.
4. Learning enough to be a fan or enthusiast.
5. Actively contributing.
6. Passing on knowledge.

Forty Suns is aimed mainly at stages 1-3—most specifically "1"—with the idea of creating new fans that are moving up the scale. Most books start in at level 2, 3 or even 4, meaning they miss the bulk of us walking around who *know* they should know more about space but who can't quite muster the interest in being interested.

We solve that.

Space is for everyone.

We should *all* be interested.

Why? Because space is our future, and the future is now.

Read on, intrepid explorer.

Let's see what we can do together.

STEM

STEM IS SHORT for Science Technology Engineering & Math. Fields that, in particular, are at the heart of humanity's collective advance into space. You might call STEM the universal language. Science is a thing any race, whether of this world or not, would speak and understand. *Forty Suns* aims to raise awareness of these universal fields, along with bringing to light the amazing advancements being pulled off in each.

Now, be patient with us as we say something Captain Obvious.

Each of those STEM fields are, in turn, made up of:

Individuals.

Duh.

But it's an important point. Strong, motivated, passionate, determined, free and free-thinking individuals are what drive our best progress.

It's those individuals we're out to support and encourage.

Which means each of us becoming more aware of what's happening in these space-related fields. Which means finding interest in the advances, the discoveries, the emerging theories that will get us there.

An interest in space begins here.

And it doesn't have to be much. As we said in answer to the question, "How does one 'Forty Suns'?", simply add it to the mix. Like a pinch of salt, it doesn't take a lot to enhance that space flavor.

If more of us were doing only that, the results would drive more change in the pursuit of space than we might think.

During the Renaissance the average man was interested in and knew much about the science and inventions of the day. He had more awareness of the world around him than many of us now do. Mostly because less was known. These days no one *could* know as much about the totality of the science and inventions of the day, as there's just too much we humans have learned in the meantime.

But each of us could know *more*.

We sit atop the shoulders of giants, who sit atop the shoulders of more giants, and on and on. Getting where we are today involved armies of STEM soldiers through the years, each pushing forward to the next evolution, the next advancement.

The daunting volume of that incredible scope of knowledge is no reason to let our own, personal knowledge lapse. And so, though most of us have become relatively oblivious (we use the technology around us but often don't understand it), *Forty Suns* is meant to be a spark to rekindle that general interest in science and invention.

Back in our day those kids who knew all the science-y stuff were the nerds. Often harassed (mostly because the harassers *didn't* know that stuff, so wedgies were a good defense against their own ignorance), in our modern, more enlightened age, we see just what rock stars those nerds truly were. In fact, when it comes to our space future, the STEM warriors of today are, in many ways, more rock star than actual rock stars.

Interestingly, the two are not mutually exclusive. Case in point: because science was once geeky and unpopular, we young lads (back when we were young) had no idea one of our rock heroes of the day was also a geek. Legendary rock guitarist Brian May, co-founder of the band Queen. There must certainly be others, but this one stands out. In addition to being a member of that culture-defining band, Brian has a Ph.D. in astrophysics and has been a consultant for NASA.

He was a STEM warrior and we never knew it.

This is but one example of the fusion of characteristics that have defined some of our many memorable figures throughout history. Told early on he couldn't do both art and science, Brian did them both anyway.

In his words:

"More and more and more, I've discovered that artistic thinking and scientific thinking are just different parts of the same thing. It's a continuum. They're inextricably linked."

It takes only a desire to explore, a drive to understand those new frontiers. Anyone can be a rock-star imaginer and creator of what comes next. STEM, like space itself (or playing the guitar), is hard. Curiosity, desire and persistence are what matter most.

Forty Suns is therefore the cure for our knowledge deficit. Or, at least, our curiosity deficit. It's the chronicle of the results of a history of scientific achievements (just about everything we talk about here can be traced through the ranks of Earth's STEM heroes), along with a way for the rest of us to become the engaged fans—or even, and especially, active participants—that encourage and drive progress.

STEM represents the pillars on the road to our space future.

Roman Locomotives

WHAT WOULD A ROMAN LOCOMOTIVE HAVE LOOKED LIKE? We'll never know for sure. But they could've been built.

Yes, that's a sharp segue, but hear us out.

From the simple to the complex, new inventions routinely come to market. We're often surprised when we see something no one ever thought of, that yet didn't technically require anything new. The Weed Eater, for example. No new technologies were involved, and suddenly we had an amazing, effective new tool. Until someone thought to combine those few basic, existing components in a novel way, we didn't have weed eaters. Putting little wheels on luggage is another, incredibly simple example. Anyone recall lugging suitcases before wheels? In each case the pieces are often not that innovative; the results, however, can be quite groundbreaking.

Similarly with the locomotive. The ancient Romans had the various bits of tech that would've been needed to lay tracks and build steam engines, hitch cars and have a great new way to move armies quickly, take trips, haul cargo and so on. No one saw that final product, and it was never invented.

Brand new discoveries are vital, but so are new ways of applying what we already know. Both are key to progress.

And so part of *Forty Suns* is the idea that, with enough interest in and focus on the fields of engineering and space science, new tools, along with new applications of existing technology, might be realized. Not only are we fans of the bleeding edge, the latest new tech, we're also looking for better ways to use what we have. In humanity's past the Dark Ages stalled and, in some cases, reversed progress. Part of the intention of this book, then, is not only to safeguard our gains (see Part Two of our mission back in Chapter I), but to promote the steady progress of existing knowledge.

If we're on the lookout for the next Roman Locomotive, we just may discover a whole new way to do things.

A bit of trivia. The seeds of this project were planted when your authors were contemplating this very thing; the idea of a Roman Locomotive, what that might've been like, and what a similar eureka-type breakthrough might look like today.

Eurekas & Creative Bursts

AT TIMES GROUDBREAKING INVENTION comes in just that way, a burst of insight, a Eureka! moment. At other times ideas come together fast in a relatively short span of time; not an instant of insight, but a rapid, cascading convergence of creativity that yields a fantastic result.

The famous SR-71 "Blackbird" is such an example.

Heard of that world-legend airplane?

Dreamed up and brought to life in 1962, with a first flight in '64 and an official introduction in '66, this plane was so ahead of its time it *still* holds the world speed record, which it set in 1976, as the fastest air-breathing manned aircraft.

The inspiration for the advanced technologies that made that possible came together in a singular moment of exceptional engineering and design (historically speaking) that delivered us one of the most amazing aircraft of this era. In fact, at this stage of our human existence the SR-71 is actually closer on the timeline of manned flight to the *invention* of aircraft than it is to us.

Hard to believe.

But if you do the math, you'll see. The first flight of an aircraft, in North Carolina by the Wright brothers, was in 1903—60 years before the SR-71. Now here we are 60+ years after *that* (as of this writing), and by that key metric—speed—we have yet to go beyond it. The Wright Flyer went about 30 mph. The SR-71 had a Mach 3.5 moment. That's creeping up on a hundred times the speed of the Flyer.

Do we have an SR-71 breakthrough looming in our near future? A convergence of ideas that yields a fantastic result?

If we keep pushing the envelope like we are, that seems quite likely.

Did You Know?

Here's a question. Did you know there have been proposals for how an actual warp drive might work? Remember we mentioned that earlier? It's not exactly a case of "Roman Locomotive", where the technologies exist and are just waiting for someone to figure out how to put them together, rather, this is an actual new thing. This incredible concept is based on sound science, and with several (steep) hurdles overcome we could pull it off.

The theory is called the Alcubierre drive. A method by which space could be warped in order to travel faster than light, the idea being that space, unlike the matter in it, doesn't have a speed limit. Lots of stuff online about it, if you know to look.

Only a theory, it has inspired many follow-on efforts which continue today. It might not even be the best ultimate solution. Likely as not it isn't. More likely we'll discover better answers along the way. But this warp drive proposal demonstrates that sort of inquisitive, confident, moonshot thinking. Miguel Alcubierre had an idea—inspired by science fiction, of course, where we do much of our future-imagining—took the physics he knew and turned it into a hypothetical for something that could actually do that amazing, impossible thing:

Cover distance faster than light can travel.

As we progress through the topics of this book we'll include these sorts of *"Did you know?"* spots along the way, giving you examples of both real and imagined technologies you might not be aware of.

If nothing else they'll give you great conversation points for Happy Hour.

More importantly such data gets you thinking with—and supporting—the things that will get us there. It raises awareness, and drives participation in the conversation.

It will take the sharing of information to accomplish our goals.

The cooler, the more interesting, intriguing, fascinating that information the better.

The Reality Facing Us

THOUGH FORTY SUNS IS DEDICATED TO PUSHING THE ENVELOPE, there's value in pausing to frame that objective.

Earth is vulnerable, and not, primarily, because of us. The planet couldn't really care less what we do. We're pretty insignificant. The Earth is vulnerable in a cosmic sense; a problem because we don't yet have other options. "Other Options" is precisely what *Forty Suns* is about. Until we get there we have a bit of a situation which, whether we like it or not, creates two major—if seemingly in conflict—mandates when it comes to our current world.

We must:

1. Take care of it.
2. Find ways off of it.

Not to escape; Earth is great. Merely to have that choice. Your home town is probably great too, but you'd hate never being able to leave. Especially if a hurricane was on the way.

Whether you yourself ultimately ever go to space isn't at the root of it. You may not even want to. But in the same way not everyone went to sea back in the day, it was important they supported and encouraged those who did (or at least didn't try to stop them). The promise of those journeys were what set us free.

The same will be true with space.

The main thing is that we develop these new technologies so that we *can* go elsewhere, when and if needed, when and if wanted. In the home town analogy, you may never want to leave, but how terrible to imagine you never could?

This isn't Chicken Little (the scared chicken who was afraid the sky was falling), or doom and gloom. We're fully behind the idea that this should be a game. That we're too amazing, too powerful not to continue to expand—and have a little fun doing it. It's merely worth noting the facts facing us, which sets the premise of the directive for constant improvement and expansion. Our playing field at the moment is inherently fragile *to our existence*, no matter how sturdy it seems.

Let's put that a different way. The Earth, this ball of rock, even by our current standards, is indestructible. Even at our current level of technology, with our bombs and our guns (did anyone just hear the refrain from *Zombie* by the Cranberries?), there's nothing we humans could do to fundamentally alter its shape. To us Earth is an impervious home.

Yet, one well-placed other-rock could change that.

And there are plenty of rocks out there.

If our planet was a basketball, the crust we live on would be like a wrapping of tin foil. If it were an apple, the land we walk and build on would be the thin skin. Pretty easy to envision that crinkled away with the right application of force, whether internal or external. What we mean to point out is that *we're* fragile, not the Earth. It's our own fragility we must safeguard against. It's never good to be that desperately dependent on anything.

These are all obvious conclusions drawn by many over the centuries. We're not saying anything new. However this is finally the period in civilized history during which we could conceivably, finally, do something about it.

Now is the time. Though many have said that before, too. For the first time in history we've reached a point where we actually have options. We could figure this out. Engineer solutions.

But we have to work for them. Make them priorities.

Together.

Space is hard. Yet we've already demonstrated, time and again, we have what it takes to do hard things.

Earth can't be our only option.

Nothing Is Making Us

NOTHING IS INSISTING WE GO TO SPACE. There is no for-sure imminent catastrophe or planet-wide imperative driving us to move fast. No timeline for disaster compels us to evacuate, or set up offworld outposts at all costs, or any other dramatic and hurried escape from Earth in order to occupy space.

That insistence has to come from within.

We must be the ones to decide that our future existence, our future civilization, our future selves depend on us taking immediate and deliberate action now.

The urgency must come from us. We must be the ones to decide those extraplanetary goals are vital, then do everything we can to make them happen.

It's up to us.

In making that decision, it should be noted, nothing says we have to trip over ourselves or panic as we go about it.

Methodical urgency, with a spirit of play, is the way.

We can pursue those goals and enjoy the game at the same time.

The most important thing is that we keep taking steps.

As such *Forty Suns* invites us to center ourselves between two extreme and diametrically opposed views, the tension of which results in positive action:

1. Nothing is serious.
2. Everything is at stake.

Let's set a good pace, see how fast we can move, and make it a point to have fun along the way.

Ad Astra

A CLASSIC PHRASE, AD ASTRA MEANS "thus one goes to the stars". That's the translation from Latin. In the space community it's used simply to mean "to the stars".

A perfect declaration of the *Forty Suns* goal.

Our mission is at once inspired by, and meant to inspire the pursuit of, that very phrase.

Ad Astra.

Can we achieve such a grand goal?

We say "yes".

But only if we cooperate toward that end.

Only if we move.

Throughout history we've demonstrated we can both help and harm each other. It's true that with increasing power comes an increasing capacity to do each other in. There's no denying most new tech is driven by the dollar sign, the demand for newer/better weapons systems, the competition to gain ground on and exceed those who aren't us, etcetera, etcetera.

Awful headlines to the contrary, we contend our human impulse, at the end of the day, is more toward help, less toward harm. We've possessed the ability to wipe ourselves out in small or large numbers for some time, and while we occasionally do so, survival and improvement always prevail.

Else we wouldn't have come forward, building the massively complex civilizations we have today.

And so, looking objectively at these facts:

1. Humanity is still here.
2. Humanity has advanced.

We must conclude that, despite any self-destructive impulses, despite any bad seeds among us, the intentions of the vast majority of us are basically good.

Though we often fall, we more often redeem ourselves, leading to positive growth. Humanity tends to correct the problems we ourselves cause. Less and less are we the effect of our environment, increasingly we're its master. With that power, and with our human faults, that control often errs, but, again, we are self-correcting, and our overall thrust is toward the betterment of our existence.

This is the premise on which we move forward.

Sure, the next world we call home may be no more utopia than this one. We might continue bad habits.

But we need that next world.

Because we'll also continue the good habits. Which far outweigh the bad. Our existence may at times be flawed, duly noted, but it's worth ensuring.

Being cynical comes easy these days. We're alive in a world that, at times, seems to foster it. We're also alive in a world that holds more promise than perhaps it ever has. Let's look honestly at both. Let's acknowledge the things that kill our hope, even as we make note of the possibility surrounding us at every turn.

Can anything be done about it?

Can we minimize the one, while focusing on the other?

When it comes to space, we believe it's vital that we do.

None of this is to suggest we have our head in the sand. Only that we choose to take the position that, with enough of a challenge to occupy us, other problems tend to fall away.

Nor are any of these opening statements meant to suggest *Forty Suns* imagines unrealistic goals—though we're definitely all about pushing for each and every next impossible frontier.

Time is of the essence.

A point to which Abraham Lincoln had this to say:

"Things may come to those who wait, but only the things left by those who hustle."

When it comes to space, at least a small majority of us must be hustling. Time, whether an illusion, a construct—whether real or not—is our greatest currency. Time is the one thing we can least afford to waste. Rapid progress is imperative.

Forty Suns is here to press that accelerator.

At once a holistic view, even as it is admittedly narrow in scope, *Forty Suns* understands that nothing works without a humanity in alignment, without a shared purpose. Our goal is that shared purpose. Which, summarized, could be stated:

Space and the furtherance of technologies that allow us to move to our next logical capabilities as a world, achieving a sustainable extraplanetary and extrasolar existence.

At the end of the day we're in this together.

What say we make it happen?

To the stars.

Checking System Pressure

HERE AT THE OUTSET, before we launch into the exciting whirlwind of space data and food for thought that is *Forty Suns*, let's pause to take a gauge of your own outlook.

We've not covered much yet. Our goal by the end is for you to experience a shift. A positive boost to your current view of our space future.

And so, when it comes to space, as you sit here reading this page, would you say you:

1. Don't think there's any chance of us ever being able to make a home anywhere but here, on Earth.
2. Think we might one day live on Mars or something, but not for a real long time.
3. See possibilities, but aren't sure if we humans can get out of our own way.
4. Are hopeful, with a pinch of optimism.
5. Believe not only can we can make space our next frontier, we will.
6. See that progress in space is happening already, and would encourage all involved—plus the rest of us—to pour the coals on the fire.

Hold onto that.

We'll check again at the end.

T-minus 3 ... 2 ...

THE ENGINES ARE SPOOLING UP. We're prepped for launch. This is the last intro chapter. Starting in the next we'll begin our full-tilt coverage of everything that goes into the pursuit of the next frontier. The first (roughly) two thirds this book will attempt to show you everything that's happening in Space Age 2.0.

A big task. You'll have to tell us how we did.

The last third, including the Appendices, will discuss ways to ensure that space future happens.

Call it two parts science and future-think, one part inspiration.

We might describe the split accordingly:

- *Chapters III through XIV* = A kind of science-y, kind of technical overview of all things space and space-related.
- *Chapters XV on* = A motivational look at ways to reach our space potential.

Chock full of goodies, we've structured *Forty Suns* a bit like a multi-course meal, each chapter intended to merge nicely with the next, leaving you satisfied yet ready for more. The next dozen chapters, III through XIV, cover the pillars of our space future:

- Power
- Space
- Habitat & Transport
- Terraforming
- Planets & Cosmology
- Physics
- Chemistry & Materials Science
- Robotics & AI
- The Human Condition
- Communications & IT
- Education
- Governance

Those are the main courses, you might say, with the final third dedicated to a variety of desserts and a few selections of port. Much like a meal, you'll spend the first two-thirds of your dining experience eating, the final third discussing life and the future while you polish that palate with higher contemplation.

Even so, the whole thing could act as a buffet. Though we recommend eating in the order presented, each chapter can, more or less, stand alone. In other words, maybe you prefer the hot buttered yeast rolls (mmm, buttered yeast rolls), but be sure to take a few green beans as well. Balance is good.

Food comparisons aside, as the clock ticks down, as we go to final ignition, one thing is certain:

The future will not just happen. We must reach for it and grab it. We must create it.

What might the future hold?

That is entirely up to us.

—

We believe pretty strongly in our future in space. You probably already get that. We'll do our best to stay off any soap boxes, but let us apologize here at the outset for any transgressions of passion. Our enthusiasm may, on occasion, bleed through.

Call us a tribe of optimists, then, when it comes to humanity's future offworld prospects. If you share a support of any of the bright possibilities outlined in this book ... well, you're part of that tribe too.

We're happy to have you.

As you'll see, you're already the hero in what may become the greatest story of our generation.

One that's only just beginning to be told.

It's been said stories are the most powerful force in our existence. Stories, throughout time, have influenced events both large and small. Our stories are what define us.

This is the story of the new space age.

May it be the prologue to a bright future.

Welcome.

It's a great time to be alive.

CHAPTER III

Power

EVERYTHING BEGINS WITH POWER. From our own bodies to the civilizations that sustain us, the world turns on the generation and consumption of power.

Power is what defines us.

Throughout most of our existence here on Earth we've not been able to bring much more power to bear than a single human could generate. More often than not that human was us. Animals became an option somewhere along the way, manual tools and techniques added to our abilities, water wheels and windmills, however humanity's extreme tech curve didn't begin until individuals were able to easily command energy well beyond their own two hands.

Once we moved past that barrier compounding advances made access to power easier and easier, in turn making the road to space a reality.

—

Global power underpins our space program and all our space initiatives, therefore in this chapter we're looking not just at space-based power, but at power across the spectrum.

The Golden Era of power began with the Industrial Revolution.

Viva La Revolution!

WHEN WE THINK OF THE INDUSTRIAL REVOLUTION we tend to think of it as a single, long event, kicking off back in the 1800s. Truth is there have been four Revolutions. The one we most recognize is actually the second, the age of mass production and electricity.

The First Industrial Revolution fell between roughly 1760 and 1840, and saw the first mechanization, steam and water power.

Prior to that not much changed. Before the introduction of those basic, power-producing sources you would've been hard pressed to note any major technological difference between eras. Sure, in one century maybe the man in the street wore bloomers and a triangle hat, in another he dressed in leather armor, but neither century had much more to offer technology-wise than the other.

Energy, power, has been the key to our advancement.

As a result, within the last two centuries we've advanced more rapidly than at any other time in human history. Not one, not two, but four stages of our "Industrial Revolution".

1.0 – 1st Revolution (1780): Mechanization, steam.
2.0 – 2nd Revolution (1870): Mass production, electricity.
3.0 – 3rd Revolution (1969): Electronics, automation.
4.0 – 4th Revolution (Now): Cyber systems.

We sit at the cusp of the 4th, which is defined as the integration of cyber and physical systems. To that, we contend, should be added the field of orbital and cislunar space (the space between here and the moon) and lunar operations.

Humanity's current efforts in the pursuit of space make up a vital part of that 4th Revolution.

Fun Thought:
Perhaps our illustrious future experts will deem Mars and a far-system human presence the markers of the 5th Revolution.

We'll see.

Whatever milestone gets set, one thing is certain:
It will depend on power.

Outputs

LET'S GET A LITTLE MORE SPECIFIC. Power, the way we talk about it here, is a somewhat general term. We use it interchangeably with energy, how we generate it, how we harness it.

To simplify, we've broken everything down into two categories:

Output
Source

Output might be called "Use", as in how we apply the power once generated or harnessed.

Source will be the energy we generate or harness, which is then used to supply those outputs.

There are three basic categories of output demand for just about all our modern human activities:

    01. Torque
    02. Electrical
    03. Thrust

There are overlaps among these, of course. As well, there are specific applications that would seem to not be covered directly in the above three. For example hydraulics, which use power to apply pressure and move mechanical systems. In that case, most likely a torque-based pump provides the pressure. Which is to say, each instance of seemingly unique application of power will have one of these three basic outputs behind it.

Torque, electricity, thrust.

Our goal is simplicity; to distill the total of *all* ways to use power down to the most fundamental.

Perhaps our future selves will need a category like Spatial Displacement (warp drives, anyone?), or something else.

For now most everything we do, from using cell phones to flying rockets, will make use of an output that fits one of the above.

Here's a quick look at each.

Torque

TWISTY-TURNY THINGS. Pretty much the way we get things moving, torque may be the OG (Original Gangsta, or "first and best") of meeting power demands. Torque can be extracted from many devices, from electric motors to gas turbines to Hemi V-8s.

Those torque-producing devices, in turn, get their power from a variety of sources (see Sources below).

Note: "torque" is actually a measure of work, not power, but it's a good word for the idea we mean to convey.

Torque is the category of power output that gets physical things turning so as to produce motion in the world.

Electrical

MUCH OF OUR ELECTRICITY COMES FROM generators, which are devices that make use of Torque. Electricity, in turn, often gets used to generate the output of torque in the form of motors. But there are other sources for electricity, such as batteries, fuel cells, etc., and there are many other outputs for electricity (the light you might be reading by, for example), so Electrical gets its own category. Electrical output is used for everything from lasers to running the heating elements (and the motor) in the dryer at the laundromat.

Acknowledged:

Overlap, as noted, is everywhere. Most electricity is used from the grid, and in most cases the grid generates that electricity by spinning massive turbines—torque-based outputs. Those, in turn, are powered by steam, which is generated by burning combustible fuels (see Sources below).

So, yes, there's no easy way to cleanly divide the entire spectrum of humanity's power supply and demand, but indulge us. For ease of understanding, these categories we've selected communicate the idea nicely.

Electricity is also where most of your renewable energy solutions fall, as most of those are harnessed to generate the electrical current which powers our world. Windmills catch wind and spin it into electricity we can use, solar cells grab sunlight, etc.

Though fossil fuels are the Source behind much of our electricity, the Outputs we rely on most around the world are electrical in nature. Even our cars are becoming more and more electric.

Did You Know?

Earth receives around 2×10^{17} watts from the sun. This is about four orders of magnitude higher than the amount presently needed to run our world. Current global energy consumption sits at $\approx 2 \times 10^{13}$ watts.

In other words, 10,000 times more energy than we need to power our entire civilization beams down on us every day.

Which means if we can get our Ready-Made Sources (see below) fully up to speed and meeting demand, we can easily power our full needs cleanly and efficiently into the future.

Thrust

THIS OUTPUT IS USED ALMOST EXCLUSIVELY for things that don't come in contact with the ground. Things that need to provide motion and action either in the air, water, or space.

Typically we use chemical-burning rockets for our missiles and to get our spacecraft into space. Jet turbines and props move our planes through the air, propellers our ships through the sea. Again there's overlap, as torque turns those turbines and props but, again, what we're looking at are final outputs.

When it comes to space, of course, rockets are the most pressing (badump-bump) application of thrust.

PUSHING THE ENVELOPE

SpaceX's Starship rocket wields the greatest thrust we've executed on yet. Seen it? Think it's huge?

It is.

Frickin huge, and almost twice as powerful as the most recent big boy to crack concrete, NASA's Space Launch System (SLS).

SLS = 9 million pounds thrust (ulp)

Starship = 17 million pounds thrust (wha ...)

Prior to that, during our last moon program (Apollo), we used the hell out of a beauty called the Saturn V, which cranked out a candle-snuffing 8 million pounds.

Not too shabby.

So yeah, we've fielded some pretty big rockets.

But get this. Way back in 1962 American rocket scientists were working up serious plans for something even more audacious. A beast that would've blown all that away, quite literally.

The Sea Dragon.

How much power, you ask?

Try 80 million pounds of thrust.

Sweet mother of ...

That's like strapping together ...

2 Starships
2 SLSs
2 Saturn Vs

... all in a single rocket. Oh, plus throw on a third Starship with 23 Raptor engines active (vs the usual 33), just for good measure. That'll get you all the way to Sea Dragon power.

This thing could never have safely launched from land.

Which is where it got the name Sea Dragon. Designed to be floated in the ocean, then set upright using ballast tanks before launch, it was affectionately called a 'big dumb booster'.

Woe be to the fish for miles in any direction around that.

But the scientists had it all figured out. The Sea Dragon was seriously considered but never implemented. The value of such a unit, able to launch 550 tons into space in one go, is not to be underestimated.

Crazy, absurd—if nothing else, it shows just what kinds of solutions we humans can come up with if challenged.

—

An Interesting Future:

Another unique form of thrust is magnetic. Magnetically levitated trains have been in operation for some time, and we'll likely soon see more advanced versions, like the Hyperloop technology being developed.

Could there be some other form of repulsive-force output around the corner? Anti-grav, perhaps?

Being able to move things through the manipulation of these invisible fields around us would be quite a feat. And quite possibly push us to the next stage of travel.

MOONSHOT MINDSET (X2)

(1) Speaking of field manipulation, prototypes for a so-called "reactionless drive" have been around for 20 years and are still

getting funding and still getting tested. This is a self-contained system that expels no mass (no flame, no burned fuel) in order to generate thrust.

Also called an EM Drive (short for electromagnetic drive), if this concept can be proven conclusively, expect development to follow rapidly. It would be an entirely emission-free, highly efficient way to move spacecraft.

A variation being tested now, using the theory of Quantized Inertia, is not technically a reactionless drive, though it provides for unique ways to move spacecraft without fuel and without violating Newton's laws of motion.

In whatever variation it eventually (if ever) finds use, this concept is bleeding-edge cool because it's definitely out there in terms of sci-fi.

(2) As of this writing NASA has awarded a grant to explore the feasibility of pellet-beam propulsion—another bleeding-edge tech that could mean the ability to accelerate spacecraft to speeds up to 300,000 mph, or 10 times faster than current chemical propulsion systems.

Definitely worth going online to learn more.

—

When it comes to thrust it's oomph vs speed. Most commonly we think of thrust with lots of kick; the kind of force needed to get a heavy rocket off the ground fast, for example. Once in space, however, and for travel over great distances in particular, the ability to provide continuous thrust for long periods is an advantage.

Ion propulsion systems are being developed for just such use. These are low-thrust drivers that nevertheless are capable of generating great speed over time.

Another example of this sort of thrust are solar sails. A better name for these intriguing craft is "light sails", as the source of their power doesn't necessarily need to be the sun. A powerful laser could do the same thing. A ground-based laser on Mars, for example, could shoot light sails back to Earth.

With about as much steady thrust as the weight of a piece of paper, and shortcomings in terms of directional control, these may nevertheless one day prove a viable intermediary solution for reaching faraway places.

At least until we get our warp drives up and running.

Sources

THERE ARE MANY SOURCES TO POWER THESE OUTPUTS. The Source, you might say, is where the true power lies, and that would be correct, however with no outputs, with no technology to harness it, a source has limited value.

We've broken Sources into four categories.

1. Batteries & Storage
2. Combustible Fuels
3. Nuclear
4. Ready-Made

As with Outputs, our future selves may have access to others, such as Vacuum Energy or some other. We'll see if future editions of *Forty Suns* (the 40th Edition, perhaps?) cover such things, but for now we're sticking with what's available, and what's on the near horizon.

Let's dive in.

Batteries and Storage

THIS IS RAPIDLY BECOMING one of our most common sources to power our lives. A battery, as a source, often represents stored energy from another Source—which means that pesky overlap we keep running into prevents absolute separation—but we feel that breaking things up this way helps to categorize and clarify our view of the energy backdrop of our existence.

Most batteries use a chemical process to release electrical energy. Our most modern batteries have a lower specific energy per unit mass than fuels such as gasoline, meaning the same weight in gas can provide more kick, but battery technologies are evolving rapidly. The higher the demand for a thing, the more rapidly it's researched and developed. Keep an eye on developments in the field of solid-state batteries, which will be the next new wave of storage tech.

The demand for batteries will continue to grow.

Did You Know?
Benjamin Franklin first used the term "battery" in 1749. He was using jar capacitors, which he grouped into what he described as a "battery", using the military term for weapons functioning together. Grouping them allowed a stronger charge to be stored, making more power available on discharge.

Many modern batteries also use multiple cells.

—

Other forms of storage exist, such as kinetic storage, where weights are raised to heights and lowered to turn generators. Hydrogen fuel cells are another way to store and release energy. See the next *Did You Know?* section below.

Note: All matter is, essentially, stored energy. Technically everything is a Source. A blade of grass, fully converted to energy, could power quite a lot. Here we're looking at the main sources we humans have figured out how to extract.

The beauty of stored energy is that it's there when you need it, with no need for conversion. In most cases (batteries, fuel cells) these storage units are also portable.

DID YOU KNOW?
There are two sources of hydrogen power that, together, span this section and the next:

1. Hydrogen fuel cells
2. Hydrogen as a combustible fuel

What's the difference between the two?

- Hydrogen fuel cells are a battery-like source.
- Hydrogen as a fuel is a combustible source and can be burned in internal combustion engines.

A fuel cell is an electrochemical cell that converts the chemical energy of a fuel (in this case hydrogen) into electricity. It does this by separating hydrogen molecules into protons and electrons. The electrons then go through an external circuit, creating a flow of electricity.

A Hydrogen Internal Combustion Engine (H2ICE), on the other hand, is an engine similar in function to the engine in your gas-powered car, that burns hydrogen instead of gas or diesel.

For either of those, how do we get the hydrogen fuel?

Hydrogen is the most abundant element in the universe, yet we still need to extract it. As a result there are three categories of extracted hydrogen:

- Grey
- Blue
- Green

Green is the clean kind, called green because, like other things we call "green", it has no CO_2 impact on the environment. Of course that means we need to pick colors for the other two.

Enter grey and blue.

Grey is for the dirtiest kind. The resulting hydrogen is green as all get-out, yet the process to extract it releases CO_2. Meaning it isn't all that clean in the end.

Blue is in-between. The CO_2 generated during the extraction process is usually disposed of somehow, as in buried or stored.

A Green Future:

Hydrogen has the potential to become a major source of clean energy. It's 100% greenhouse-gas free, it's clear, colorless, odorless, plus it has the highest energy density of all chemical fuels. A definite win-win.

The biggest challenge, as noted, is extracting it. Mainly because we've so far not developed ways to find and recover it from naturally existing sources.

This is changing.

Natural, or "geological", hydrogen is now on the radar of private and national enterprises, and it may be more widely available than was supposed. In the same way finding plentiful oil in the ground was once dismissed as a crazy idea, finding hydrogen locked up in the earth may become just as routine.

We all know what happened with black gold.

The new "gold energy" may be clear, and it may be coming soon.

Combustible Fuels

OUR OLDEST AND MOST COMMON SOURCE. The 1st Industrial Revolution saw the burning of combustibles (Source) to make steam, which was used to power steam engines (Output). From there better fuels developed, our technologies with them.

In fact, when it comes to the harnessing of energy, combustible fuels go back even further. Fire warmed our bodies and cooked our food, all the way back to the first time we learned how to make it.

As for space, combustible fuels' major role will be in powering the rockets that get us there.

For our Earth-based engines, cleaner, alternative combustibles exist or are in various stages of development, such as:

- Biodiesel
- Ethanol
- Natural Gas
- Propane
- Renewable Diesel

Plus the aforementioned hydrogen.

So much of our current infrastructure is powered by internal combustion engines that burn fuel on demand, we'll likely be using them for some time to come. Alternative and emerging fuel sources are ways to reduce the impact of their emissions.

DID YOU KNOW?

Methalox, or methane-liquid oxygen, is a fuel combination used to power rocket engines that run on methane and liquid oxygen. SpaceX's Raptor series of thrusters use this fuel source. As new engine designs prioritize reusability, as well as what's known as in-situ resource utilization (ISRU), or using resources at the destination to replenish—as with missions to Mars—this combination of methane and oxygen is becoming the standard for next-generation launch vehicles.

While human occupation of Mars is still in our future, methalox fuel is favored by Mars exploration advocates, as both methane and oxygen can be sourced locally on the Red Planet.

Plus methalox rockets burn that cool, translucent purple-blue.

They just *look* like science fiction.

Nuclear Fission

WITH NUCLEAR ENERGY WE MOVE CLOSER TO A CLEAN, pure solution; one which provides loads of sustainable power. There's a huge amount of energy locked up in the nucleus of each atom.

At present we harness nuclear energy by splitting atoms in a reactor to heat water into steam, turn a turbine and generate electricity. This is the process called fission, and it's the one we've been doing for decades around the world. Land-based powerplants and sea-based reactors (aircraft carriers, submarines, cruisers) use the same process.

As of this writing the Inflation Reduction Act (IRA) in America has made nuclear fission eligible for the same tax credits as renewables like wind and solar—which is quite likely to herald a fresh nuclear renaissance. After years of decline in the industry we could soon see a resurgence. In fact a recent report by the DOE (Department of Energy) suggests America could triple its land-based nuke power in the next 15 years.

Nuclear is a clean, compact way to deliver huge amounts of power to the grid. Expect to see lots of it coming back online.

DID YOU KNOW?
Companies like NuScale out of Oregon are designing a new breed of small reactor known as a Small Modular Reactor (SMR), a class of nuclear fission reactor which can be built in one location, then shipped, commissioned, and operated at a separate site.

The modular nature of SMRs provides wide flexibility. Additional units can be incrementally added as needed, and new units can be rapidly produced at a decreasing cost following the completion of the first reactor on site.

Another company, X-energy out of Rockville, Maryland, is building gas-cooled reactors that pioneer the use of a special fuel that cannot melt down.

As these SMRs become more safe, smaller applications could find their way into use. Applications such as emissions-free powerplants in seaborne craft (that would probably be smart; shipping accounts for 3% of global emissions), to say nothing of even more advanced applications in freight locomotives and, imagine, over-the-road trucks. Personal vehicles? Likely not—we'll come up with better ways—but nuclearizing our freight transport infrastructure would be a significant step toward cleaner solutions.

Look for an increasingly fission future, at least in the near term, as more of these advanced reactors and SMRs are brought online and put into use.

Nuclear Fusion

THE HOLY GRAIL OF NUCLEAR POWER IS FUSION. This is where atoms are fused rather than split, releasing much more of their potential energy. Fusion takes a huge amount of heat. In fact, so far it takes more energy to actually produce and control the fusion than we get out of it. More power is required to run the magnets, the lasers, etc. to start and sustain the reaction than is produced.

But we've done it. We've created fusion reactors and fired off fusion reactions. The resulting plasma is hotter than the core of the sun, is turbulent and hard to manage, releases high-energy neutrons during the process that bombard the metal walls of the reactor, can produce beams of high-energy electrons that bore holes in the reaction-chamber cladding (yikes), along with other challenges that crop up when you try to ignite, create and control a star on land, but we're doing it.

Wild stuff.

As with all new technologies, it's only a matter of time.

Did You Know?
A fusion project called ITER, Latin for "the way" and an acronym for International Thermonuclear Experimental Reactor, is one

promising gambit. Not only for the technology it will pilot, but for the players involved. ITER represents a veritable who's-who of global participants.

Parties involved include:

- The European Union
- The UK
- China
- India
- Japan
- South Korea
- Russia
- The United States

This is the biggest, and most expensive, such undertaking, requiring the sorts of resources only a consortium of national players like this can bring to bear.

But talk about important.

Expense, of course, being relative. For perspective, take a look at this comparison of Cost vs Result:

Cost of ITER = $12B
Result = clean, pure energy to run the entire world

Cost of Vietnam War = $120B (10x as much as ITER)
Result = ... ?

(Yes, it's okay to pause for a moment. The comparison is as insane as it seems.)

The difference in value proposition is clear.

And did you notice the who's-who involved in this little project? Aren't some of those guys competitors in other arenas?

Indeed they are.

Another joint project, between the European Union and Japan, the JT-60SA reactor, has gone live. A collaborative precursor to ITER, the six-story-high machine in Naka, north of Tokyo, uses a donut-shaped "tokamak" vessel to contain the plasma.

The beauty of these sorts of grand, collectively-beneficial efforts is that not only do they have the potential to advance humanity, they have the potential to become Great Unifiers.

Space Age 2.0 is just such common ground.

DID YOU KNOW?
The Tokamak shape used in most experimental fusion reactors is a torus (donut) shape, wherein the plasma is confined in a symmetrical ring. Another shape gaining popularity is the Stellerator. Also a donut shape, the stellerator puts the plasma through a twisting, asymmetrical motion.

Both use magnetic confinement to handle the super-hot plasma, and each has certain advantages. While tokamaks are better at keeping plasmas hot, stellarators are better at keeping them stable.

Though tokamaks are more prevalent at the moment, stellarators could become the option of choice for our first fleet of fusion reactors.

PUSHING THE ENVELOPE
Fusion requires us, in essence, to create a star on Earth, but we're doing it. Novatron Fusion Group, out of Stockholm, is at the cusp of a major breakthrough, which will help bring this dream to life.

Their NOVATRON fusion reactor design is an innovative solution for stable magnetic plasma confinement, intended to be a steady-state continuous process with high reliability, providing stable plasma confinement with high energy density.

All signs point to a commercially viable reactor in the next decade. The first is already being commissioned and developed. Target is the 2030s—which is practically just around the corner as these things go.

Let's keep an eye on this one.

—

When it comes to fusion, commercialization, as with all things, will make it viable. Though fusion is an expensive undertaking, it's not all about big national and international projects. As of this writing a few dozen startups are active around the world, at least three of which may have what it takes to bring this incredible technology not only to life but to market.

Novatron Fusion Group above being among them.

On the nation-state front, in the US the National Ignition Facility (NIF) has, more than once now, squeezed enough energy out of a diamond capsule packed with hydrogen to keep the fusion reaction running. Called "ignition", this is when the energy released is enough to sustain the fusion process.

Wherever and however it first achieves viability, fusion will very likely become the major Source of power for our next generation. No matter what other advances we make in the near term, controlled fusion is the future of global—and space—power.

PUSHING THE ENVELOPE (X 2)

(1) One great space-related application of nuclear power being worked on is the nuclear rocket. Called a Nuclear Thermal Rocket, or NTR, these promise to be a next-gen solution for covering the vast distances in space. As of this writing NASA and DARPA (Defense Advanced Research Projects Agency) have begun a new program to develop this exciting method of space travel.

In their words:

"In a nuclear thermal rocket engine, a fission reactor is used to generate extremely high temperatures. The engine transfers the heat produced by the reactor to a liquid propellant, which is expanded and exhausted through a nozzle to propel the spacecraft. Nuclear thermal rockets can be three or more times more efficient than conventional chemical propulsion."

NASA has said they want "robust and enduring access throughout the solar system." NTR will likely be part of realizing that mandate. In fact they've just chosen a contractor to begin the design and testing phase.

(2) An actual *fusion* rocket is also in the works. Rather than using a nuclear reaction to heat fuel, this one is what's known as a Direct Fusion Drive (DFD), where charged particles create thrust directly. It's powered by atomic isotopes, and doesn't need a huge fuel payload.

Such a powerplant could potentially drive spacecraft at speeds of up to 500,000 (half a million!) miles per hour.

Hummanah.

The CEO of the company developing this new tech, Pulsar Fusion, had this rather practical observation to make:

"You've got to ask yourself, can humanity do fusion? If we can't, then all of this is irrelevant. If we can—and we can—then fusion propulsion is totally inevitable. It's irresistible to the human evolution of space."

We like that viewpoint.
Fusion is our future.

Moonshot Mindset

As we've noted, fusion is what powers stars. In an object the size of a sun the collapsing gravity of its mass provides all the pressure and temperature needed to fuse atoms. When we do it on Earth we have to make that pressure with extreme heat.

What if you could do it at room temperature?

In effect, releasing that same energy bound up in the atom, but doing so in a controlled way without the heat and temperatures required otherwise?

Cold Fusion holds the promise of such a feat.

At present this is only a hypothesized type of nuclear reaction. The idea of Cold Fusion is extreme, nevertheless due to its promise it has been attempted many times, and even evaluated twice by the DOE (Department of Energy).

In general this sort of next-level pursuit is exactly what science (at least some of our science) should be about.

—

Antimatter is even further off, but if we could generate or store anti-matter particles for propulsion or power generation we could release the entire energy stored in each particle. Smashing together particles and their anti-particles would, therefore, make for an incredible energy yield per kilogram of fuel.

For fun, here's the comparative energy yield per method:

- Fission = 0.1% mass converted to energy
- Fusion = 0.4-0.7% mass converted to energy
- Antimatter = 100% mass converted to energy

Meaning antimatter would yield 200 times more energy per unit mass "ignited" than fusion. And fusion is already ultra bad-ass.

For a baseline, fusion generates around 4 million times more energy than a comparable mass of fossil fuel. Put another way, if a gallon of gas will get you 30 miles, that same amount of material fused would get you 120 million miles, or 100,000 trips back and forth between New York and Miami.

That's an insane MPG rating.

Converting that gallon fully (as with antimatter) would get you 20 million trips. Yeah. Basically, a car that would never, ever run out of gas.

—

Harnessing the power available at the core of an atom is the ultimate Source at present. We can expect increasing development of this, in particular fusion. Variations of atomic yield will become a large part of our future space infrastructure.

Ready-Made

RATHER THAN CONVERTING MATTER of some kind to energy, Ready-Made sources already exist as energy in one form or another. Also called sustainable, natural, renewable, or 'green', this is the energy all around us. Example, solar. In that case we're merely capturing one form of energy (heat, radiation) and converting it to another form we can use (electricity).

All energy sources are "natural", in that they come from nature at some level. The advantage of Ready-Made sources is that they represent energy that's already there, without needing to burn, fuse, or otherwise bring into existence energy that was locked up in another form.

This is, of course, a potentially great answer to most of our power needs, yet one which needs refinement. How do we capture that energy? Wind farms are unsightly, mar the countryside, are less green than anticipated and inefficient. Solar farms can't fully feed the machine—but they're getting better. Cost-effectiveness must be addressed. Wave capture devices haven't really panned out. Others are being imagined and tried.

As with batteries, demand for solutions is high. Progress in this area should continue at a healthy pace.

—

At one of Tesla's recent investor events, a plan was outlined on the premise that a sustainable energy economy is within reach, and we should bend our will toward accelerating it. With that goal as a common purpose, setting our minds toward achieving the amazing steps outlined, that future can be—remarkably easily—achieved, with big gains for humanity up and down the scale.

Closing the event Elon Musk had this to say:

"This is not just about the investors who own stock, but really anyone who is an investor in Earth. What we're trying to convey is a message of hope and optimism. Optimism that is based on actual physics and real calculations, not wishful thinking. Earth can and will move to a sustainable energy economy, and will do so in your lifetime."

Honestly, this is one that's worth looking up. A sustainable energy economy is within reach. They've done the research, run the numbers, and laid out the projections for an actual way we can reach that major milestone.

This is the type of forward thinking we should be supporting.

Unnecessary-necessary Note:
Vested interests will always slow such efforts. Existing systems that are difficult to undo. Arguments about infrastructure and the money people need to make on consumables they're counting on us to keep burning so they can sell more.

That sort of resistance should never stand in our way.

Moonshot Mindset (x 2)

(1) Geothermal energy is ready-made, but sometimes tricky to find the right conditions in order to tap. The core of our planet is hotter than the surface of the sun—all we have to do is drill deep enough to liberate some of that heat.

How do we do that?

One way is with light. Instead of drill-bits we'd use beams of light to vaporize rock. These contactless drills could bore holes as deep as 12 miles into the Earth's lava-hot crust. That heat could then be used to convert water to supercritical steam, which would then drive standard turbines and produce electricity.

A big energy solution could be right beneath our feet.

(2) Then there's the possibility of beamed energy. Space Based Solar Power, where the sun's energy would be captured by satellites in space and beamed down to Earth, has been around as a concept since the 1960s. Recently several nations have launched projects to make it a reality. The advantage is that these stations would collect sunlight energy at all times. There would be no cloudy days or night. As envisioned these power stations would sit in geostationary orbit (see next chapter, *Space*), collect sunlight, convert it to energy and beam it to receiving stations on Earth using microwaves.

In this case, our energy solutions would be right over our heads.

Fun Fact:

Based on current levels of growth, the commercial space sector is projected to reach a total value of $1.4 trillion by 2030. Interestingly, the ready-made energy sector is projected to reach the same level, $1.4 trillion, before the end of the decade.

Power As A Yardstick

WHEN IT COMES TO POWER THERE'S A SCALE, called the Kardashev scale, scientists use to classify the advancement of civilizations.

Everything begins and ends with power, therefore any civilization can be rated according to how much power it commands. The Kardashev scale is broken into three very broad Types that span a wide range. A Type I civilization, for example, is able to use the energy available on its entire planet. A Type II that of its entire star.

Here are the three basic Civilization Types, ranked according to the scale:

- **Type I** = a civilization that can harness all the energy that reaches its home planet from its parent star.
- **Type II** = a civilization capable of harnessing and channeling the entire radiation output of its star. This is the entire star, not just the portion of energy reaching the planet.
- **Type III** = a civilization having access to power comparable to the luminosity of the entire galaxy. Heady stuff.

Earth is not yet a Type I. Physicist and futurist Michio Kaku suggests that, if humans increase their energy consumption at an average rate of 3 percent each year, they may attain Type I status in 100–200 years, Type II status in a few thousand, and Type III status in 100,000 to a million years.

We believe a concerted focus on the exploration and colonization of space will drive that rise much faster.

Humanity has certainly exceeded expectations before.

No matter our rate of expansion, Power will be the yardstick by which that advancement is measured.

The Illusion Of Scarcity

WHICH BRINGS UP A REALITY that must be both understood and widely accepted. When it comes to power "more" is the goal. Our future demands it. Humanity's power needs will continue to increase and that's a good thing. Increasing demand for power means we're doing more, which means we're growing as a civilization. We'll find more and better ways to make it and, as we do—advancing up the Kardashev scale mentioned above—we humans will eventually get on the map as an actual extraplanetary civilization.

That won't happen if we focus on ways to conserve.

Our goal is an abundance of power.
Not less.
Clean, reliable, sustainable, yes. Of course.
But gobs and gobs of it.
An extraplanetary existence depends on it.

Sam Altman, visionary behind much of the recent AI hullaballoo (see that upcoming chapter), is backing a parallel push on both AI *and* better sources of energy.

He had this to say:

"The alternative to not having enough energy is that crazy de-growth stuff people talk about. We really don't want that. I think it's insane and pretty immoral when people start calling for that."

He was referring to a philosophy of restricting production, consumption and energy use as a way to conserve natural resources. A flawed way of thinking. Certainly no way to build a huge, expanding, ambitious future.

There is no scarcity except that which we impose on ourselves, or agree to settle for. It helps none of us to think in terms of less. The idea of a "lack" of things is a short-term illusion—apparently real in a very limited sense, but not actually true.

We sit at the doorstep of a vast, vast universe (as we'll see in the next chapter) with unlimited everything. One we'll never reach if we throttle back.

So let's not go down that road of thinking small.
Let's think big instead.

—

Whatever humanity's power needs, for whatever purpose we apply that power, our goal with *Forty Suns* is how that power can best be directed toward one end:
Space.

CHAPTER IV

Space

"SPACE, THE FINAL FRONTIER ..." One of the most iconic, and exciting, openings, firing the imagination of would-be space explorers everywhere.

Who doesn't recall that trippy theremin intro? (If you don't remember the theremin, the instrument that makes the distinctive *Star Trek* opening sounds, it's worth checking out for a shot of nostalgia.) *Star Trek* is definitely a solid example of what a far-future existence might be like.

But while Kirk and crew had the power to boldly go, the reality is, living and moving in space brings with it a whole new set of challenges.

Big ones.

In *Star Trek* they had a lot of those licked. In the same way that, compared to a cavemen, we've got living and moving around easily licked on modern Earth (a caveman would be stunned by how fast we can go from New York to London—or that there even *is* a New York and a London), Starfleet had space figured out in ways we can only imagine.

Yet, mastery of a thing doesn't mean the challenges are no longer there. It just means we've come up with ways to deal with them.

Did we mention space is hard?

Forces

MANY FORCES VIE TO END US in the unforgiving environs of space. We prefer a little gravity to give us a sense of direction, for example. Oh, and to hold us to things. Air for breathing is also nice, along with the pressure of an atmosphere.

Radiation can be a bummer. Temperature, while not a force, is also something we need to moderate. Not too hot, not too cold.

Humans are pretty frail. That's just being honest. Admitting that let's us know where we need to begin. Don't get us wrong. It's fun being human. Being human is a great way to experience the world. But we need an awful lot to be *just* right in order to stay alive, let alone have fun. In planetary science we speak of a Goldilocks zone for water to exist, and therefore life as we know it. Being human requires an even tighter zone than that. It's not enough that water simply doesn't boil or freeze. To exist in our native form we need a Goldilocks zone where, not only is the temperature of the porridge just right, it's got just the right oat count, the precise amount of locally-produced honey, it isn't too lumpy, it's non-GMO, gluten free, and has no refined sugar.

A lot has had to go our way for us to even be here.

Put another way, when it comes to living off Earth even the baddest of the bad among us are no match. Heck, even *on* Earth we need things to be just-so in order to live. A SEAL Team Six survivalist might put the average person to shame when it comes to making it a week in the Klondike naked and afraid; he's just like the rest of us when it comes to the moon.

In short, to survive we need help.

Fortunately we're great at making stuff. And when it comes to space, we've already come up with a ton of great ways to keep us not only alive, but more or less fully functional.

Survival

MAKING USE OF EVERYTHING FROM ANIMAL SKINS to jackets to spacesuits we've been coming up with ways to help our bodies survive for a great, long time. Whatever the frontier, whether exploring high on a mountain, under the ocean, in the Arctic or in space, we meet the challenge of each new extreme with innovation.

There's an old saying, "There is no bad weather, only bad gear."

We humans learn quickly how to adapt and survive.

When it comes to space, we've now been out there a bunch. Both locally, in orbit, and a few manned trips to the moon. In all that we've figured out quite a bit about living and surviving in this wholly unforgiving environ.

Sections later in this book cover considerations of human physiology, habitation, and more.

Mastery of space, in short, is well within our grasp.

DID YOU KNOW?

A private company has secured the contract to make the space suits that will be used by astronauts making the next landing on the moon.

Artemis III (we'll talk Artemis more later) is the third mission in the much-hyped and super-important Artemis series of missions to return humans to the moon and prepare the way for human missions to Mars. Axiom Space, one of the many incredible private companies focused on our space future, will provide the new Axiom Extravehicular Mobility Unit (AxEMU) spacesuit that makes that possible.

When astronauts return to the Moon for the first time in over 50 years, they'll be wearing these next-generation spacesuits to walk on the lunar surface.

Pretty cool, eh?

Fun Fact (because we know you're wondering):

We've all heard statements on what would happen if you were exposed to space without a suit. Believe it or not you'd only last about 15 seconds. Your blood holds enough oxygen for that much brain activity, which means after that you'd black out, with complete brain death following within three minutes. Death by asphyxiation, would be the coroner's evaluation.

Distances and Travel

NEXT CHALLENGE: EVERYTHING IN SPACE IS FAR.

A quote from Douglas Adams in *The Hitchhiker's Guide To The Galaxy* says it best:

"Space is big. Really big. You just won't believe how vastly hugely mind-bogglingly big it is. I mean, you may think it's a long way down the road to the chemist, but that's just peanuts to space."

Truer words.

Analogies often help when imagining things we don't have direct experience with. The subject of distance in space invites silly ones, like how many credit cards it would take laid end to end to reach the moon. These references get unwieldy quick, but that's mainly because there *is* no good way to compare it. The idea of 5 billion credit cards laid end to end is about as inconceivable as the distance itself.

Space is it's own unique thing when it comes to scale.

Time helps. Framing distances in times at least makes it real to our human brains how far away a thing is. "It would take a rocket as long to fly to Mars as it would take you to run and swim around the entire world." Another silly analogy, to be sure, but at least it's one we can simply reduce to a period of time—7 to 9 months, nonstop. Distance it ain't, but it gives us a way to conceptualize it. Which is about the best we can hope for when it comes to travel in space. Probably why we use a time measure when talking distances to other stars. A light year (the distance light travels in a year) makes it way easier to express.

(Though, again, conceptualizing the distance light travels in a year is, well, mind-bogglingly difficult.)

Point is, those vast distances involved in space travel *really* get in the way of making things easy. It takes a *lot* to get anywhere once we leave this world.

Yep. Definitely sounds like a challenge.

Which is why we humans, being the big-thinkers, action-takers and crazy-dreamers we are, have not only done it, we're gearing up to do lots more.

Moving In 3D

WHEREAS THE SORT OF TRAVEL WE'RE USED TO here on Earth typically involves moving in two dimensions, space requires more calculations. Convenient reference points are often lacking. In addition to missing control forces, such as gravity or air, we'll be moving in all directions.

We'll continue to rely on instruments and computers for even the most basic maneuvering. Getting around will involve more finesse and methodical execution than a human can provide.

Meaning seat-of-the-pants space jockeys flying spaceships like *Star Wars* X-wings may be in our future (we sure hope so), but not with this generation of technology.

DID YOU KNOW?

Speaking of this sort of movement, expect the demands for better on-orbit maneuverability to continue to, um, accelerate. The days of simply putting things in orbit and making only minor adjustments as they mostly just hurtle around the planet following the laws of inertia will, more and more, be behind us.

Recently, in its latest Hyperspace Challenge accelerator, the US Space Force selected three startups specializing in satellite propulsion. Their objective? Nimble satellites that can maneuver to advantage over adversaries.

We'll be seeing more of this.

On-orbit dynamism will soon become a requirement. Many state players are fielding increasingly maneuverable craft; a sign that this is the future. Officials with the USSF are calling for this shift to "dynamic space operations."

It's a logical progression. We expect these technologies will, eventually, after a few (several?) evolutions, lead to those first X-wing fighters.

PUSHING THE ENVELOPE

In lower gravity scenarios, such as on the moon or on Mars, devices called mass drivers could be used to get things into orbit. This is a proposed method of non-rocket space launch that involves the sequential firing of a row of electromagnets to accelerate a payload along a path, catapulting it out the end at high speeds.

Think of it like a really big, really precisely tuned gun.

A mass driver on Earth would most likely be used to accelerate a payload up to some high speed, releasing the payload, which would then complete the launch with rockets.

Mass drivers use coilguns rather than railguns, making them less of a cannon/gun, more of a rapid accelerator. The sequential and timed nature of the electromagnets allows velocity and acceleration to be more finely controlled.

That way your cargo doesn't get squished.

Note:

Companies like Auriga Space, part of the new space economy, are already planning such novel electromagnetic launch systems for use right here on Earth; ground-based electromagnetic tracks that accelerate launch payloads at speed, using a similar technology to a high-speed maglev train.

As mentioned, the electromagnetic launch system takes the place of a first-stage rocket to reach altitude, eliminating the need for a concussive, high-emission launch. A second stage rocket is then used to reach final orbit.

Space Age 2.0

WHEN IT COMES TO OUR EXTRAPLANETARY FUTURE it helps to organize our objectives. Space, after all, is the focus of this entire book; our key subject matter.

What might our space battle-plan look like?

For starters (and that's really where we're at; this is the start of the second Space Age), we might organize our key objectives into categories like this:

- Space Transportation
- Space Exploration
- Space Governance
- Space Infrastructure
    - Earth Observation
    - Telecommunications
    - Navigation
    - Support Services & Logistics

These are the areas in which we can group the space-based activities that will catapult humanity to the next frontier.

What Are Some Realistic Objectives?

INCREASING OUR ORBITAL MASTERY, through ongoing payload launches to continually perfect those systems. That would be one objective. Housekeeping of our current assets and debris, another. Ordering and coordination of missions, along with expanding the presence of real live humans within major orbital spheres:

- LEO (Low Earth Orbit) | < 1200 mi
- MEO (Medium Earth Orbit) | 6K – 12K mi
- GEO (Geostationary Orbit) | 22K mi

Including Cislunar (between the earth and moon).

More and more spaceborne bases (see next chapter, *Habitat & Transport*), many of which are planned or on the way, dedicated to various tasks and built for various purposes, will, as well, help increase this mastery and presence.

A permanent lunar occupation, of course.

Naming our first true Big Objective, of many to come.

To really get out there we need to *be* out there.

By having a base and an orbital station at the moon, we make it much easier to:

- Launch larger missions deeper into the solar system.
- Gather and collect resources we'll need for living in space.
- Test and develop the space-based systems that will be part of our expansion.
- Acclimate to living in a space environment.

The moon is perfect because it's just a few days from home, with a less than 3-second delay for radio communication.

Lunar & The Artemis Program

SPEAKING OF THE MOON ... WHAT'S UP? As mentioned, Artemis is NASA's next-gen program to get us back there. With elements planned in cooperation with the ESA (European Space Agency), Canada and Japan, this new push for the moon will be nearly too big to fail. Exactly the threshold we want to cross. In many ways the far-thinking Artemis program epitomizes and centers Space Age 2.0.

NASA's Chief Exploration Scientist, Jacob Bleacher, said:

"This (the Artemis program) is turning the first page on a brand-new chapter of space exploration."

NASA even has a tagline: "Rebooting the Moon."
If that's not a hallmark of Space Age 2.0, we don't know what is.
Learn more at NASA.gov, under Missions.
And NASA isn't the only one. Many interests, both private and national, have their eye on the moon.

Having such a lunar presence—call it a way-station—is an ideal way to establish ourselves in space, while we figure out the details of how we reach toward Mars and beyond.

Beyond that, three primary factors have been driving fresh interest in the moon:

1. Confirmation that water-ice is likely to exist at the poles, as well as in permanently shadowed craters. The presence of water solves many problems for a lunar presence and exploration. We need it, and if we can find it, and don't have to haul it then, Hey, that's great.
2. The rise of China's space program. Competition, as we know, is one of the best drivers of progress.
3. Interest from private companies in the commercial development of the lunar surface.

That last one in particular is a great sign of our progress. Corporate investment in the moon, from companies like Astrobotic, Intuitive Machines, Firefly, ispace (a private Japanese company) and others, is the sort of thing that will build the foundations of a permanent extraplanetary presence.

If we keep this pace, in a few years, when you gaze up at our closest celestial neighbor, pale white in the day, shining bright in the night, your fellow humans will be living and working up there.

How cool will that be?

We're damn close to that being a fact of life.

In many ways the moon, as NASA's Mr. Bleacher implies, is the first step toward our space future.

PUSHING THE ENVELOPE

One of the most well-known corporate enterprises pushing our space future is SpaceX. Their Starship program has many purposes, one of which is to provide transport for upcoming lunar missions. Talk about a sci-fi future. If you haven't checked out the Starship program by SpaceX, we invite you to do so.

Beyond that, Starship's ultimate goal is Mars.

Occupying Mars will take huge resources and focus, but with companies like SpaceX setting and going for this goal, others are soon to follow. As with most things, all it will take is one success, followed by economic opportunity, and Mars will more and more be a part of our everyday lives.

In the objective of achieving a "second home" option for humanity, Mars is our nearest, best solution.

Orders Of Magnitude

SPEAKING OF THOSE MARS AND LUNAR objectives, it's helpful to take a look at how our current operations need to scale. In order to achieve a meaningful offworld presence fundamental numbers are in play; orders of magnitude increases to what we're doing—and what we're capable of doing. These numbers are where the rubber meets the road, so to speak. Or, in this case, where the LOX (Liquid Oxygen) hits the rocket fuel.

What might be needed to really make space an everyday reality?

The total mass of objects in space right now is roughly 8,000 tons. That's everything that's been launched since 1957. Nothing to sneeze at. A big American freight locomotive weighs in at about 200 tons. Imagine the energy it took to throw 40 of those suckers all the way up into orbit. Launch them so fast and so hard that they actually stayed up there.

That's what we've been doing, as a world, for the last sixty years.

So it can be done.

Now. Let's consider what's needed.

Estimates for an ongoing Moon and Mars presence, with related operations, could take sending from 1 to 5 million tons into space. Let alone orbital interests. Bigger better space stations will also "weigh" in with their needs (badump bump), adding to the overall.

Those are big numbers.

To give it a real-world reference, the Empire State Building is about a third of a million tons. Which means sending 1 to 5 million tons into orbit would be like firing 3 to 15 Empire State Buildings over the horizon.

In addition to being fun to watch, you can imagine how much energy that would take.

Split the difference and that's a little over 300x the mass of what's up there now. Which means, at the rate we've been launching stuff over the last 60 years, it would take around 18,000 years to make a multi-million ton target.

Clearly we've got to step up the pace.

Best way to do that is with massively bigger launch systems, coupled with more frequent launches. SpaceX's Starship is what we might consider the first generation of the next generation of what will be our massive-lift options. Starship, at full burn, can snatch the weight of one of the aforementioned locomotives.

Yep.

Let's pause for a moment on that one.

Next time you're standing near a big freight train, stop a moment to really absorb what you're looking at. How *heavy* it is. Touch it if you can. Maybe break off a piece of iron between thumb and forefinger, just to get a feel for that much mass. (Kidding; unless you're The Hulk you probably can't do that.) Now imagine what it would take to hoist one of those—nay, *hurl* it—fast enough and high enough to go into orbit.

Starship can do that.

Hm.

Bringing it a little closer to the present, Falcon Heavy, current workhorse and heavy-lift champion in the SpaceX fleet, can hoist over 60 tons. That's not a locomotive, but it's about one MBT, or Main Battle Tank. Send up an American M1 Abrams, or a South Korean K2 Black Panther, or an Israeli Merkava, or a German Leopard 2—your choice. Book a few flights on the Falcon, send up one of each and have a tank battle in space.

That might be kind of cool, actually. Imagine the recoil dynamics; tanks madly pinwheeling in all directions.

Of course that would *really* add to the space junk problem.

But we digress. A little.

Here's where it relates. By the numbers, Starship could get a comparable 8,000 tons into orbit (what we have up there now) in 40 goes. That 1 to 5 million ton target? Call it 12,000 launches at max load.

If we wanted to do that in a year we'd need a fleet of mighty Starships thundering skyward 30+ times a day, every day, including National Donut Day.

Yes, space waits for no holiday.

At this stage those numbers are general and speculative, but they're in an expected range, which points to one absolute certainty:

We've got to elevate our game.

Estimation Of Effort

TO SECURE OUR SPACE FUTURE we need to be prepared to push harder, move faster, think bigger. Disrupt our tendency to stare inward; put an end to our willingness to accept a methodical pace. We must instead be willing to engage upon mighty bouts of adventurism.

It's important. And it will be worth it.

As is said, the rising tide lifts all boats. If we're gaining in space, we're gaining across the board. There may be lags. Gaps in results may occur between success in one arena and success in another. When we have our first base on the moon, for example, it doesn't mean other world problems will be instantly solved.

But it's a guarantee they will be.

With big gains come the resources and the knowledge to solve not just the current problems, but bigger problems still.

(That also works in reverse. It's a guarantee that if we don't expand outward we'll collapse inward.)

Forward, not back. Always attaining, always achieving, always a new win, a new goal ahead.

At this stage of human existence we're too close, too much has been done already, to do otherwise.

Our estimation of effort must be in line with the future we demand.

Corporate Influence

GOVERNMENT SPACE PROGRAMS and commercial ventures are pursuing the development of methods for that magical in-situ resource utilization (ISRU) we mentioned earlier. ISRU is the local mining and processing of the resources needed for a remote base to be self-sustaining (whether moon, Mars or otherwise).

Once ISRU is up and running it won't be long before surplus resources will become available, and those same mining and processing actions will begin to provide materials—and even products—for us right here on Earth.

That will be a huge milestone.

Space Mining

BEING ABLE TO MINE RESOURCES OFFWORLD will drive the next fervent wave of interest in space; the very impetus needed to kick us humans to the next level of conquest. (Building ships in space will be another big milestone. Doing that, refueling in orbit ... these things will dramatically change the time and frequencies we can expect from interplanetary travel.)

Mining water, metals and oxygen from the Moon and asteroids is where it will start. Water-ice on the Moon could provide:

- Potable water for bases and settlements.
- Oxygen for those same installations.
- Rocket fuel.

Plus helium-3 and rare-earth elements, along with other minerals that might be found in the lunar regolith (the rocky surface layer), which contains things like aluminum, iron, titanium and silicon. Asteroids, as well, contain all sorts of valuable metals, including platinum, nickel and iron. Any of these could be used as materials for building in our space-based environs, or brought back to Earth for use here.

Once there's profit to be made and supply chains to stuff, we'll enter a new era of accessibility and resource wealth.

Absurd To The Nth Degree:

As we go to press the Psyche probe has been launched, a mission meant to investigate 16 Psyche, an all-metal asteroid worth an estimated $100 quintillion dollars.

A hundred what?

Annual global GDP (the output of the entire world) is around 100 trillion, meaning that asteroid, by the numbers, is worth a million times what the entirety of us produce in a year. Meaning a person in possession of such an asteroid and its resources could, in theory, own planet Earth—at its current output—for the next ten thousand centuries.

Take that, Powerball.

Of course that's absurd, on many levels, but it does brilliantly illuminate one very important point:

There's a *lot* of stuff out there.

We have only to get to it.

Did You Know?

When it comes to "mining", we just witnessed an example of an actual mining expedition in space. With the OSIRIS-REx mission NASA actually:

- Sent a probe into deep space to rendezvous with an asteroid (asteroid Bennu).
- Navigated said probe to make contact with said asteroid (a 500-meter target) and deployed a robotic arm.
- Used the arm to scoop up samples from the asteroid.
- Flew the probe right back here to Earth.

Mad skills, yo.

Not only that, JAXA (Japan) has already had *two* such missions; Hayabusa 1, which collected materials from asteroid Itokawa and returned them in 2010, and Hayabusa 2, which returned samples of asteroid Ryugu in 2020. What's cool about OSIRIS-REx, though, is that it's still going. It dropped its cargo and is now headed off on the next mission: the asteroid Apophis. Like a true space mining truck, it went, it mined, it collected, it returned, it delivered, and now it's off for more. Not a new craft; the same one, continuing to the next assignment.

Sounds like science fiction, and maybe it once was.

Like most things, however, these sorts of missions—and operations in space in general—will one day just be regular old science reality.

Commercial Space

RIGHTLY OR WRONGLY PROFITS, as noted, will be a prime motivator. Altruistic visions for space notwithstanding, money is and will be the impetus for the speed of our expansion and the development of new technologies. The above examples are primarily scientific in nature, but eventual space mining will need to make financial sense.

Profit and capitalism drive our world. From wars to medicine to space races, things advance because there are gains to be had. It doesn't make the process bad. Neither does it make it perfect. Capital is simply the reality of our existence and, knowing that, we can use it, like anything, to our advantage.

As international corporate interest in space grows so, too, will our opportunities for profit and, as an extension of that, our occupation of space itself.

There's even a new term for it:

NewSpace

Official definition:

Aerospace companies working to develop low-cost access to space or spaceflight technologies and advocates of low-cost spaceflight technology and policy.

Government no longer needs to intervene entirely. These NewSpace companies are now able to operate autonomously in many cases, an independence which is only increasing.

Expect commercialized space to become a major pillar of our space future.

A Wild Idea:

If we continue pushing our mastery of the space domain at the rate we are, it won't be long before we begin sourcing more and more of our resources offworld.

Who knows? One day might we get *everything* we need from "out there"?

Such a state would mean no more digging for stuff here (except maybe for dinosaur bones and buried pirate treasure), meaning we'd eventually be able to leave our home planet as pristine as possible.

Yet another advantage to opening up the space frontier.

To Infinity And Beyond

AND WHAT ABOUT STILL FURTHER GOALS?

How about *much* further?

Breakthrough Starshot, a project of the Breakthrough Initiative, is a research and engineering program which aims to demonstrate proof-of-concept for a new technology, enabling ultra-light uncrewed space flight at 20% of the speed of light. This will lay the foundations for a flyby mission to Alpha Centauri within a generation.

Alpha Centauri is the nearest star system, and is just 4 light years away.

Such an enterprise could be our first visit to an actual other star.

PUSHING THE ENVELOPE

In a more local sense, solar system-wise, the fastest human-made object ever built is picking up speed. In 2024 NASA's Parker Solar Probe will reach a sphincter-pinching 430,000 miles per hour.

That's one trip around the Earth's equator every three minutes plus a few seconds or so.

Each time it slings around the sun it gets faster, meaning in the next year it will continue to break records (its own). As of right now the Parker Probe is only in competition with itself. In addition to traveling 150 times faster than a really fast rifle bullet, the probe is doing other things, like gathering valuable data on the sun and, extra cool (literally), showcasing heat shield technology which has allowed it to pass through our local star's outer atmosphere.

Of course the speed has been the splashiest headline.

Nour Raouafi, an astrophysicist at the Johns Hopkins Applied Physics Laboratory—and project scientist for the mission—confirmed the significance of that record-breaking velocity by making the following, highly scientific observation. Though most of us reading this aren't scientists, we'll go ahead and include his technical assessment here.

Mister Raouafi said:

"It's very fast."

—

Of course, in the case of a star shot, there's the potentially comical possibility that, once such a long-range flight is launched, we make great strides in propulsion science back here at home. Imagine a *Eureka!* Moment where we develop warp drive, way ahead of schedule, then send a new craft to the same destination, only to be waiting there when the slower one arrives.

Unlikely, but possible. In fact anything is possible, we like to believe, which is the whole idea behind stoking the fires of interest in space. That wide interest, that curiosity, then drives all the other important decisions that need to be made among the world's research and development efforts, thus pushing us faster toward mastery of the next frontier.

A final comment on the road ahead:

In every great thing we've ever accomplished, "hard" had to happen before we could have "easy". When it comes to that truism, we've mentioned a few times (only a few?) how challenging space is.

Let's dedicate the final section of this chapter on Space to that. With, of course, a positive twist.

That is, after all, what we're all about.

(Imagine Tony Little energetically encouraging us all, "Yeah, baby! You can do it!")

Space Is Hard

WE'RE LIVING IN A WORLD OF LEISURE. Built and maintained, of course, by lots of hard work.

The problem is we tend to settle into the easy, even as, in our comfort, we sometimes forget to do the hard. Or even that we need to. But we do. In order to have the next great "easy" thing, we need to get to work.

It's okay to enjoy the easy; we earned it. Long term, however, complacency kills. Failure to set bold goals, and take bold action (often the missing piece; setting goals is easy, putting them into play is hard), will see us wither.

There's an old saying:

"Hard times create strong men, strong men create good times, good times create weak men, weak men create hard times."

That cycle is a real one. Yet cycles, vicious or otherwise, can be broken. There's nothing that says we have to sit around waiting for the next hard time in order to get motivated and get stuff done.

We can start anytime we choose.

In a famous speech that, in many ways, launched the first Space Age, JFK invoked this concept:

"We choose to go to the moon in this decade and do the other things, not because they are easy, but because they are hard."

There can be little doubt of where that motivational speech led, and the results it achieved.

We're on our way again, with the advent of Space Age 2.0. It won't be any easier this time. It might even be harder.

Are we up for it?

One of NASA's slogans is "Failure is not an option". Only, how is that possible? Our national space program has definitely had failures.

That phrase was coined during the Apollo 13 mission, when the lives of three astronauts were at stake and absolutely everything that could be done was being done to bring them home safe. During the tension of that harrowing return flight, failure, indeed, was not an option.

As a more enduring slogan, however, how could there possibly be no room for failure? After all, the only people who don't make mistakes are the ones who don't do anything. So ... if we're doing stuff, aren't we bound to make mistakes?

Aren't we bound to fail?

Ask yourself "failure when?" and you have the answer.

The expectation is not to fully eliminate failure. Failure is in the fiber of every successful enterprise. You could even go so far as to say that in order for an enterprise to succeed it must have failed. Attempting anything, especially anything hard, brings with it a near certainty of failures.

What NASA means by failure not being an option is that failing at the *goal* is not an option. It might take many failures to get to the finish line, but the goal, the objective, *will* be attained.

And this is how we win.

DID YOU KNOW?

Most space missions are measured in terms of the number of "single points of failure". In other words, how many chances are there for the entire mission to fail that hinge on one thing going wrong?

The recent JWST (James Webb Space Telescope) mission had the highest potential single points of failure for any mission in NASA's history—even Apollo.

Yet, with over 300 ways JWST could fail utterly with just one thing going wrong (300!) ... nothing did, and we now have one of the most incredible scientific instruments in our history.

This sort of defying the odds is what we humans are all about.

It's how we enjoy the advances we do today.

More than that, we're here because many of us have been willing to take that calculated risk.

While in the case of JWST the overall mission didn't fail, you can bet that if it had we'd continue to chase the goal. Which is why, in the end, we succeed.

Failure is not an option.

In each case of accomplishing the incredible we see examples like this, and not just in space; examples of multiple obstacles being overcome, of large groups of people with both strengths and weaknesses working side by side, people with differences and commonalities, coming together, fighting against all the reasons why it shouldn't work until they ...

Succeed.

—

Space is hard, yes.

But so has been every other monumental success throughout human history.

So no worries.

We humans have a habit of pulling off the incredible.

CHAPTER V
Habitat & Transport

OKAY. SO SURVIVING IN SPACE AIN'T EASY. As noted, perhaps unnecessarily, we're not exactly made for it. And that's simply surviving, let alone figuring out ways to actually be comfortable. The thing that *does* work to our advantage, as also noted, is that we're pretty damn good at making tools and coming up with ingenious solutions.

Look around here on Earth, and it's clear we've figured out all kinds of ways to exist—and thrive—in difficult environments. We regulate temperature everywhere we go, from our clothing to the buildings we live in. Those same solutions shield us from the elements. We've got the whole food thing pretty much nailed (though food distribution is another matter; another way the unfriendly among us exert control). We can protect ourselves from any threats the wild might have to offer. We've made it super easy to move these bodies from place to place. Sure, we can walk or run if we choose, but if you really want to get somewhere just pick your solution.

Planes, trains and automobiles are standing by.

Space is the next frontier, and we've already figured out a lot in that direction.

Surviving and thriving in space will call upon some of our greatest tool-making skills and most ingenious solutions yet, but good news for our space future:

We've convincingly shown we're up to the task.

Orbital Operations

SINCE 1971, 12 space stations have been launched into Low Earth Orbit (LEO) and have been occupied for varying lengths of time.

Here's the list so far:

- 01. Salyut 1   (1971)
- 02. Skylab   (1973)
- 03. Salyut 3   (1974)
- 04. Salyut 4   (1974)
- 05. Salyut 5   (1975)
- 06. Salyut 6   (1977)
- 07. Salyut 7   (1982)
- 08. Mir   (1986)
- 09. Tiangong-1   (2011)
- 10. Tiangong-2   (2016)

Rounding out the dirty dozen, as of 2023 there are two fully operational space stations up there:

- 11. The International Space Station (ISS)
- 12. China's Tiangong Space Station (TSS)

Most of humanity's space stations have been fielded by Russia, a few notable entries by the US and a consortium of countries (the International Space Station), and, most recently, China.

Private industry will be next.

Planned private stations and dates include:

- Axiom Station   (2025)
- Starlab Space Station   (2027)
- Orbital Reef Station   (late 2020s)

Those are in the works, with others in the wings.

And let's not forget the Big Daddy, the best one so far, the Lunar Gateway, planned as part of the Artemis (moon) project, being executed by another consortium: the US, the ESA (European Space Agency), Japan and Canada. It's slated for "soon".

That's the one intended to serve as a science platform and as a staging area for the lunar landings of NASA's Artemis program, along with follow-on human missions to Mars.

Living and working in space is about to become rather routine.

Which is exactly what we want.

PUSHING THE ENVELOPE

NASA has announced they're working with SpaceX to make SpaceX Starships into Space Stations. The SpaceX Starship has more volume than the International Space Station, and could hold a significant, working crew.

Some of the design renderings of the inside compartment usage look not only modern and productive, but pretty dang cozy.

MOONSHOT MINDSET (X3)

(1) What about a ring station?

We first saw one of the best fictional examples of such a station in 1968's *2001: A Space Odyssey*. Also called a Rotating Wheel, or von Braun wheel, this design provides an artificial sense of gravity around the outer ring by rotating at a fixed velocity. The resulting centrifugal acceleration holds occupants and items to the floor of the outer ring. Designs have been imagined or put forth, and such a thing would be within our near-term capabilities.

As with most new stuff, what it would take would be a collective decision and the allocation of brain power and resources. NASA has had its reasons for not yet attempting this.

Namely:

- It would be hard to construct, given the current lifting capability available to the US and other spacefaring nations. As we see in the chapters of this book, that's changing.
- Assembling and pressurizing such a station would bring with it many obstacles—and expenses. Not impossible, just hard. And pricy. Again, our willingness to see the value in such things is also changing.
- The current station, the ISS, has been meeting demands.

Could a private company find reason to construct such an orbital presence? Maybe a space hotel?

From what we hear Vast Space has been giving this a genuine look.

(2) Taking the ring-station concept up a few notches, another grand idea is something called an O'Neill Cylinder.

Proposed by American physicist Gerard K. O'Neill, an O'Neill cylinder would make use of materials extracted from the Moon and later from asteroids, and would consist of two counter-rotating cylinders (to counter gyroscopic effects), each 5 miles wide and 20 miles long. Habitation would be within the cylinders.

There have been other such proposals from scientists as well as from several masters of science fiction. Sci-Fi examples of this have been seen in shows like *Mobile Suit Gundam*.

According to Dr. O'Neill's vision, life aboard would be better than some places on Earth. An abundance of food, along with climate and weather control, continuance of pastimes, such as sailing, skiing, mountain climbing, hiking and camping—plus all new pastimes that might evolve, making creative use of the fact that everything exists within a massive cylinder—would add up to a paradise-like existence.

Even more grand than a space elevator in terms of engineering required (see the *Getting Around* section below), this sort of concept would nevertheless be a feasible order-of-magnitude project for a global human effort. The technology would be within our grasp.

(3) Fringe planetary options. With a bit of extreme ingenuity, we might craft habitats on a few worlds that otherwise seem impossible to inhabit. Venus, a miserable hellscape (see the next chapter on Terraforming), *does* have certain regions of the atmosphere where we could actually walk around in our civilian clothes. We'd need air to breath, of course, but otherwise the conditions would be bearable. At about 30 miles above the surface the atmospheric pressure is very similar to Earth, with temperatures hovering at a balmy 80 degreed Fahrenheit. Imagine some sort of zeppelin-style cloud city, levitated by a nitrogen or oxygen combination, which are lighter than the predominant carbon dioxide atmosphere.

Sounds pretty sci-fi groovy.

Plants could even be grown.

Similarly, Saturn's big moon, Titan, while cold and bleak, does have bearable surface pressures and could be survived with the right cold protection gear and, of course, breathing equipment. Titan is bigger than Pluto and Mercury, and has quite a few uncanny Earth-like features.

Could Saturn's largest moon become prime real estate for a new breed of hearty solar system settler?

Sustainability

BEING SELF-SUFFICIENT will be a hallmark of any otherworld installations. Part of planning for all future bases includes figuring out ways for the occupants to replenish supplies, gather raw materials, perform construction and repair, etc. Resupply, as you can imagine, is not as easy as placing an order on Amazon.

Space-based habitats don't have much to draw from. Recycling for them becomes key. In the case of the ISS, for example, most of the station's oxygen comes from electrolysis, which uses electricity from the ISS solar panels to split water into hydrogen gas and oxygen gas.

As for water, the ISS recycles much of its water using chemicals, but still relies on sizeable shipments of water from Earth to give the astronauts access to clean agua.

Surface habitation, as in a Moon or Mars base, could make use of resources available there.

Did You Know?
You've heard of companies launching rockets and satellites, but, in case you missed it in that list up there, there are private companies actively planning to build actual *space stations*. Permanent presences in orbit around our Big Blue Marble, built by private enterprise.

In fact NASA has partnered with several that will be helping with various mission objectives, even other than those listed above. With the International Space Station (ISS) scheduled to retire in 2030, NASA is placing a huge emphasis on a seamless shift to future private space stations in low-Earth orbit.

Recently, as of this writing, the Aerospace Safety Advisory Panel has called on NASA to provide a comprehensive understanding of the requirements needed to transition from the ISS to commercial space stations, called Commercial Low-Earth-Orbit (LEO) Destinations, or CLDs.

"NASA should develop a comprehensive understanding of the resources and timelines of the ISS-to-commercial-LEO transition plan to a much higher level of fidelity, to provide confidence that the nation will be able to sustain a continuous human presence in LEO," said David West, a member of the panel.

NASA and private industry are all over this one.

Starting now, living and working in space will no longer be exclusively the domain of state players.

PUSHING THE ENVELOPE (LITERALLY)
A small-scale prototype of Sierra Space's Large Integrated Flexible Environment (LIFE) module was deliberately blown to pieces after spending a month withstanding high pressures well above what will be required of it in space.

Yes, it's an inflatable space station.

Part of the Orbital Reef space station, a small set of industry-led space stations that will replace the aging ISS, this module by Sierra Space is a real example of pushing the envelope—quite literally.

Living & Working

WHEN IT COMES TO LIVING AND WORKING OFF-EARTH, the next biggy will be the moon. The Lunar Gateway project (mentioned above) will provide a spaceborne station in orbit near the moon, which will act as a way-station (a gateway) for future activity, but there will also be bases *on* the moon.

As part of that Artemis project, NASA says:

"We will build an Artemis Base Camp on the surface and the Gateway in lunar orbit. These elements will allow our robots and astronauts to explore more and conduct more science than ever before."

Eventually the station will allow astronauts to spend up to two months on the lunar surface.

All this will also prep us for developing the next-gen habitations that will found our increasing extraplanetary presence.

Mars, anyone?

DID YOU KNOW?
Already we've conducted thorough experiments simulating living and working on Mars. The latest is kicking off at NASA's Johnson Space Center in Houston; the Crew Health Performance

Exploration Analog, or CHAPEA for short, will explore more of what we'll need to know when we send our first inhabitants to the Red Planet.

Mission 1 is already underway.

When it comes to the real deal, robots will likely go ahead of us, and accompany us, constructing our future off-world habitats in advance, then helping us to run them when we're there.

Advance preparations for human occupation, as with the CHAPEA experiment, are vital.

—

Our first lunar base will start small, facilitating missions of a week or two, but as the camp grows in size and sophistication longer two-month stretches will be realized. Plans call for a lunar cabin (like a camping cabin; a place to live and from which to take excursions), plus an open-top rover similar to the kind used in the Apollo missions. Plans also call for a larger vehicle, something like an RV that would provide mobility while allowing astronauts to live and work away from the base for days or weeks at a time. Like leaving the log cabin and taking a lunar road trip.

The idea of all this being to use those complimentary elements not only to collect resources, but to create a base from which we can develop and refine techniques for remote habitation, which then act as a jumping off point to go deeper into the solar system.

Sounds like the beginnings of a TV show some of us may remember from the 70s, *Space 1999*.

Very cool stuff.

Getting Around

ONCE THE ROCKETS GET US THERE, what next?

How do we move ourselves, and our things, around in space, or in low-gee environments?

And here's a crazy question:

What might we use to get us there besides rockets?
Read on.

Zero Gee, No Surface

SATELLITES AND EVEN SPACE STATIONS have positional rockets for maneuvering. Hydrazine thrusters are used to perform small corrective maneuvers and attitude control on large spacecraft, and may be appropriate to act as the main propulsion system for smaller ones. For satellites, typically 3 thrusters are adequate to control attitude and provide translational thrust.

Note:
Keep an eye on nitros oxide (just like the stuff used in the *Fast & Furious* movies). Nitros-based systems are growing and are suitable for all sorts of mission applications. What's more, N_2O is readily available with a well-established supply chain. Look for this to become a new standard in 0g maneuvering.

On-orbit specialized vehicles are also being developed or are on the way. We have so many things in orbit now, companies are devising ways to provide services like refueling, removal and placement.

Think space tugs.

A space tug is a type of spacecraft used to transfer spaceborne cargo from one orbit to another. An example would be moving a spacecraft from a low Earth orbit (LEO) to a higher-energy orbit like a geostationary transfer orbit, a lunar transfer, or an escape trajectory.

As a side note, tugs could be used to gather and dispose of space junk. This would be helpful as, in addition to the active things we maintain in orbit, there's a lot of trash that's been left behind.

Tidying up space, plus reaching agreements that reduce future junk-piling, will be a big part of our space future.

For perspective, here's what's still hurtling around up there:

(Number of spent rocket bodies and other pieces of debris.)

- Russia   7,032
- USA   5,216
- China   3,854
- France   520
- Japan   117
- India   114
- ESA   60
- UK   1

We'll see more and better ways to maneuver in space as we continue to expand in that arena.

Pushing the Envelope
Startup company, iRocket (Innovative Rocket Technologies), won a Space Force contract. IRocket has a proprietary afterburning rocket engine that will be key in building a reusable small launch vehicle able to lift 300 to 1,500 kilograms to orbit.

This will be a smaller-use example of reusable propulsion systems. Stand by for more and more of these as Space Age 2.0 heats up, bringing the cost of doing business in space lower and lower.

Low Gee, Surface & Air

Depending on the surface, different sorts of specialized vehicles will be designed, from jump jets to wheeled or tracked vehicles. How we get around will depend on the application.

When it comes to the moon, the lunar terrain vehicle (LTV) will be the first big step toward establishing a base camp. It's scheduled to arrive on a mission sometime after Artemis III in 2025.

Check out some of the GM and Lockheed Martin LTV design concepts. We want one!

Advances in battery technology (see the *Power* chapter), along with motors and drive components (including locomotive arms, either electric or hydraulic), and the lightweight yet strong materials we'll make them out of, will play vital roles. We've already seen our distant autonomous exploratory rovers last much longer than expected, using long-term low-yield nuclear generators. Advances of similar technologies will help when it's us out there, riding around on (no doubt very cool looking) Martian dune buggies.

And what if there's an atmosphere?

Even if not much of one, we've now demonstrated the ability to fly on another planet.

The Mars Helicopter, *Ingenuity*, flew for the first time in April of 2021. A very Earth-centric technology, adapted for an alien world, this little helicopter flew around and maneuvered in the thin Martian air much like our helicopters and drones of Earth. In January of 2024 a rotor blade broke off and other blade tips were damaged during the landing, but talk about a proof-of-concept.

Ingenuity. An appropriate name. A perfect example of the very human ingenuity that made it happen.

Next up: *Dragonfly*, a helicopter mission planned for Saturn's moon, Titan.

Moonshot Mindset

One way to get things into orbit isn't with rockets, but with something way more normal-sounding:

An elevator.

Design specs for such an orbital delivery method have been worked out in detail. More as a thought exercise at this stage, but the concept is promising enough it's captured the attention of many a great mind.

Our engineers and scientists know exactly what it would take; it's just a matter of the sheer cost of the effort, and the engineering feats involved. Kind of like a version of building the pyramids for us modern humans. About as many engineering miracles (relatively) and as much focus of manpower would be involved.

But, like the pyramids, it could be done.

The idea is essentially a long cable extending from our planet's surface into space, with its center of mass at geostationary Earth orbit (GEO), or 35,786 km in altitude. A simple concept, amazing engineering and materials tech required.

Even NASA calls space elevators "audacious and outrageous".

Yet they've done extensive studies.

With a concerted global effort, we could make one.

Pushing the Envelope

When it comes to getting around, hypersonic propulsion is heating up and will start to see commercial use soon. We might employ it in other atmospheres as well.

Hypersonic flight is flight through the atmosphere below altitudes of about 50 miles at speeds greater than Mach 5. Several companies are working to develop this technology for reliable use.

Check out Hermeus (hermeus.com) for a pretty spectacular example.

SOPs For An Extraplanetary Existence

ALL WE NEED TO MAKE AN EXTRAPLANETARY EXISTENCE happen are a few new SOPs (Standard Operating Procedures). Guides by which to conduct ourselves, such as:

1. Focus our collective attention on these extraplanetary, moonshot goals.
2. Set our sights on larger and larger ambitions, pushing the envelope at every turn.
3. Develop, test, refine the ways we'll habitate and move around.

Our mission, should we choose to accept it, would be to adopt such standards, make them part of our everyday existence, begin to operate with this mindset and soon enough our options outside Earth would begin to multiply—even as they come to feel a lot like home.

Sounds like exactly what we're after.

—

But wait!

There's more.

Cool as all this is, the *real* ultimate score would be having a world other than our own where conditions were much the same. Not just comfortable bubbles, underground caverns or space-rings we could occupy, but an actual, livable land surface we could walk and live on. An "Earth 2" where we could stroll country lanes in the open air, picnic in the fields and so on.

A dreamy proposition. And not a small undertaking.

Begging the question:

Is such a thing possible?

Could we make a new Earth?

CHAPTER VI

Terraforming

EARTH IS PERFECT. Or nearly so. Of course our world is perfectly suited for us humans; it's been our home since ... well, since as long as we humanoids have been living here. We're quite symbiotic with our chosen planet.

How great would it be to be able to create these same living conditions elsewhere, on a planetary scale?

That's the idea behind terraforming.

"Forming" an existing planetary body to be like Earth (terra) is what it's all about, and, yes, it's about as big of a thing as it sounds.

But there are viable theories for ways it could be done.

—

Though the mere idea of terraforming is, admittedly, epic, even to consider, we decided it's important enough not only to include it, but to give it its own chapter.

Terraforming, whether a few generations from now or much further down the road, is in our future.

Mars

EACH NEW WORLD WE WANT TO TERRAFORM will take its own, unique methods. Mars, for example, would most likely be done in stages, first by releasing elements locked up in its rocks and in its surface, thus creating a thicker, oxygen-rich atmosphere.

One way to do that would be with lasers. (In our mind we're placing a pinky finger to one side of our mouth, like Dr. Evil, as we say 'lasers'.)

Lasers firing from space would melt the surface, heating and releasing the elements needed to build up the atmosphere. There are strong indications that Mars once had an atmosphere as thick as Earth's, during an earlier stage in its development, and that the pressure of that atmosphere supported abundant liquid water at the surface.

Odds are very good we could create those conditions again.

After that we'd take it to the next level.

There are multiple steps that would have to happen, but essentially terraforming Mars would break down to building up and tailoring the:

1. Atmosphere
2. Biosphere
3. Magnetosphere

That final piece, the magnetosphere, would be needed to shield the results of our hard work. Turns out the atmosphere Mars once had was probably slowly blasted away by our life-giving yet oh-so-rude sun, because the planet doesn't generate a magnetic field that could've shielded it.

We could solve that by placing a magnetic dipole field between the planet and the Sun to protect it from high-energy solar particles. It would be located at the Mars Lagrange orbit L_1, which is a stationary orbit that would hold the magnetic shield in position between Mars and the Sun at all times.

One other technique might involve simply placing such a shield to begin with, which would allow Mars to partly restore its own atmosphere.

Venus

CALLED EARTH'S "EVIL TWIN", Venus, believe it or not, would be an even better candidate—though the atmospheric conditions there are super extreme in the other direction. The air on Mars is thin and arid; Venus makes the Amazon basin look like, well, Mars.

Thick, hot ... not pleasant.

Still, Venus is a whole lot more like Earth.

For starters the size and mass are really close. Gravity on Venus is 90% that on Earth, which would make an eventual terraformed Venus way more familiar to us when walking around, carrying bags of groceries, etc. Long stays in low-gravity will have negative health effects on our human bodies, which is an issue if we're going to decide to live our lives elsewhere.

For comparison, Mars has just 35% of Earth's gravity (0.35g).

When it comes to making a planet Earth-like, Venus is a tougher nut to crack, but the payoff could be a world a lot more like home than Mars.

The three main steps for terraforming Venus would be:

1. Reducing the surface temperature.
2. Eliminating most of the planet's dense carbon dioxide and sulfur dioxide atmosphere via removal or conversion to some other form.
3. The addition of breathable oxygen.

Interestingly, the first phases of this would likely involve freezing the CO_2 out of the atmosphere, meaning that, with Venus being about as close an analogy for Hell that exists, in a sense, to begin fixing Venus we'd need to make Hell freeze over.

Final factors. Venus rotates way slower than Earth. One Venus day is 116 Earth days, which means that during each "day" cycle that side bakes brutally in the sun. We'd either need to change the rotational speed (a mind boggling amount of energy needed to do that, let alone the technique—but not impossible), or fabricate an in-space mirror to regulate the day/night cycle.

That second one is likely the route we'd take.

After that we could tailor the last of the atmosphere and begin adding life.

Note:
When we say Venus would be better than Mars, we're strictly talking a terraformed Venus. At this moment in our history, with the technology we have and will have in the near future, Mars is the way to go. In its current state Venus would be way inhospitable to any habitation solution we could dream up.

The Red Planet is where we will live next.

DID YOU KNOW?
We're already *kind of* taking terraforming actions right here on Earth.

Cloud seeding is a weather modification technique that improves a cloud's ability to produce rain or snow by introducing tiny ice nuclei into certain types of subfreezing clouds.

It's not new, and it's one way we're already influencing outcomes within massive atmospheric systems. In this case our own.

Closer To Home

ON THAT NOTE, reverse engineering our climate right here on Earth would be another application of these thought processes. Things like CO_2 removal systems are already being developed and put into use. Many other ideas have been put forth.

Whether you subscribe to these ideas or not, the fact is we need to clean up our act. Though the Earth has moved naturally through atmospheric cycles far more extreme than any currently experienced, we've now reached a critical stage of human development. One where our civilizations require near perfect balance. Meaning engineering climate stability is more important than it's ever been.

We, personally, can take the extra hot or cold.

Our modern civilization can't.

Ten thousand years ago if the sea rose due to a hot swing, we'd just pull up stakes and go make a bunch of new straw huts further inland. Back when we were hunter-gatherers, if glaciers started freezing us out we simply migrated south.

No biggie.

Now we're rooted. Now our massive infrastructure and our very way of life require an incredibly fine-tuned environment. Too much change can, literally, ruin us. Compared to our ancient selves we're now frighteningly inflexible. Our tremendously complex and precarious global civilizations can't adapt the way we, with our simpler lives, once could.

And so, since we can no longer adapt to the environment, we must work to adapt the environment to us.

Which means preserving its stability.

Terraforming-type techniques may be one way to go.

Taking a really broad view, if we have it in us to think through the crazy science it would take to turn a Venus into an Earth, surely we can take an already near-perfect solution (Earth itself) and tweak it enough to make it perfect again, or at least hold it steady.

Something to consider.

One Day

TERRAFORMING IS A HUGE STEP UP from our current capabilities, but with one of our most major goals as humanity being to create (or find) other places to live, thinking with the idea of altering these two most suitable local candidates is a good use of our time.

Mars and Venus would be our first, logical objectives.

Each would be multi-generational projects, but then many of our most significant accomplishments throughout history have been. Heck, the cathedral of Notre Dame took two centuries to finish. We were willing to put in the time and effort on that, why not a planet?

In the case of the cathedral (or many such large-scale projects), for us to enjoy them past generations had to work their asses off knowing they'd never see their work completed.

Which highlights an optimum survival balance for any successful civilization. That is, being willing to take steps not just for oneself, but being able to look ahead and take steps for future generations as well. Actions you take now from which you yourself may never directly benefit.

We may never live on Venus, but if we keep moving the ball forward someone down the line will.

And that's what it's all about.

One day similar steps could, and likely will be, taken for planets around distant stars. Each new world, of course, requiring its own, unique solution.

There are great videos and articles online if you'd like to learn more about terraforming.

—

Okay. Beyond Mars, beyond Venus, how do we find those next planets to terraform?

Or possibly even worlds we can occupy as-is?

Wouldn't that be neat.

Keep reading. The journey continues ...

CHAPTER VII
Planets & Cosmology

IN ANCIENT TIMES the solar system was thought to be the entire Universe. Through successive discoveries we've learned just how vast our Universe truly is.

And it may only be one of many.

There's (currently) a horizon to what we can see. Even in our modern age we talk of an "observable" universe. Everything we can take in. Meaning it's quite likely there's more of it out there beyond the limit of what we're able to view. How much more ... well, with the speed of light as a limiter we can only see as far as the universe has been expanding.

Which is roughly 13.7 billion years.

Anything beyond that is unknown to us.

And 13.7 billion light years isn't even the actual distance of the things we see. You'd think that would be the size, but what we're seeing is further because, since the universe has been expanding all that time, the light we see now, coming from that distant edge, has been traveling our way nearly 13 billion years.

Which means during that time that light was racing *to* us, the things that emitted that light have been racing *away* from us. Those distant points have continued to fly toward infinity.

Which means ... the universe has grown in those 13 billion years. Rushing away from us, even as the light we see now rushed toward us. And so those galaxies we see in our telescopes at the edge of the observable universe are, in fact, currently much *much* further away than 13 billion light years.

And, of course, a lot, lot older now.

Based on calculations of the expansion of space (remember, space doesn't have a speed limit), the observable universe we're seeing today is actually now more than 46 billion light-years in any direction. That makes the whole thing, with us at the center, about 93 billion light-years in diameter.

Anything beyond that we, with our current technology, can never know. Meaning that "edge" we see is not the actual edge.

And, of course, we're not actually at the center. Only at the center of what we can see. Meaning ...

What was it Mr. Adams said about space?

"Vastly hugely mind-bogglingly big."

Lest you think we're being mean, we're not. Our purpose in any of this is not to make you feel small.

Actually, quite the opposite.

Our purpose is to give you an appreciation of just how *much* possibility is out there. Space, and all things in space, could be summed up under one heading:

Opportunity.

Humanity, in a sense, has only just begun to explore.

—

Intelligent Aliens & Other Civilizations:

Let's pause a moment to address one of the most interesting things about space: The idea that other smart, civilization-building beings might be out there. The thought of this is hugely fascinating, if for nothing else than pure human curiosity.

Are we alone?

Odds are incredibly, ridiculously good that we aren't.

In fact, stretching our minds a bit, if the universe *is* infinite or nearly so, or if there *are* more universes besides this one, the odds are not small that there's another version of you out there somewhere, living pretty much the same life you're living now. Only today they're wearing red shoes and you're wearing white.

You get the idea.

Point is, nearly anything is possible.

Bringing it back home a bit, in a more local, galactic sense, if there are other advanced civilizations alive today, especially spacefaring ones, one of the biggest questions in modern cosmology is, Where are they? Why have we found no record?

That may change any day now, but for the moment we've found no confirmed markers of alien intelligence.

This is a full subject unto itself, as you may imagine, and worth a little research of your own. Fascinating stuff, with lots of very smart postulates floating around by lots of very smart people as to what the mystery of the seeming absence of advanced other-beings might mean.

We won't get into it here—at least for now. Our focus is on us humans and what we can achieve all by our lonesome.

If others start showing up, especially others smarter than us, well, that will no doubt change things.

Perhaps a future edition of *Forty Suns* will address this.

Planet Hunters

IN 1992 WE DISCOVERED the first ever extrasolar planet. It actually ended up being two planets; named Poltergeist and Phobetor, found orbiting a pulsar called PSR B1257+12. Later, in 1994, a much smaller 3rd planet was found in the same system, named Draugr.

Since that first discovery in 1992 we haven't looked back.

Over 5,000 exoplanets and counting, many of which are of a size and type that could host life; potentially habitable worlds occupying that magical Goldilocks zone around their host star.

Two hundred so far that are earth-like

55 of which would quite likely be bearable for us humans.

55?

Yep.

That's already 15 more than our arbitrary 40 suns goal.

And what about the rest? The ones in "the zone" that might be too big, too gassy (eww), tidally locked, or otherwise non-optimum for life. Do we automatically discount them?

What about their moons?

What if they have moons large enough to harbor atmospheres and life? Moons around planets in the habitable Goldilocks zone might beat the issues plaguing their host, providing a viable combination for life.

New methods of detection are being worked on each day. Either imagined, planned, in development, or slated to go live. With the new James Webb Space Telescope we're scanning distant planets to determine atmospheric composition, and therefore whether or not those planets might support life.

New telescopes will expand the search.

In fact, we've also just recently demonstrated the ability to detect a magnetic field around one of these distant worlds (YZ Ceti b), a near requirement for life to form. A good strong magnetic field helps deflect solar winds which might otherwise stunt the development of life, as well as erode a planet's atmosphere.

Remember what happened to Mars?

Soon we'll even be able to detect signs of life. In the near future we'll even find ways to scan for the markers of *intelligent* life.

Imagine the day we get *that* news.

Such is the march of science. One by one each mystery becomes known.

One by one each new discovery becomes part of our scientific knowledge.

Pushing the Envelope

James Webb Space Telescope, or JWST, was launched Christmas of 2021. Not only is it 'out there' in terms of what it can do, it's literally out there—nearly a million miles from Earth. We sent it all the way out to what's called a halo orbit, circling around a point in space known as the Sun–Earth L_2 Lagrange point. This keeps it in line with the Earth, such that Earth is always between the telescope and the sun. Great for better views of the heavens, great for staying in communication with our armies of scientists here on the ground sending it all kinds of instructions and receiving all kinds of data.

The JWST has been making milestones since it came online. As of this printing it's shown us:

- Stars being born.
- The most distant galaxies ever.

- Spectral readouts of exoplanet atmospheres.
- Titan's clouds (moon of Saturn).
- Stellar details never before seen.
- Hidden star formations.
- A direct *image* of an exoplanet.

You read that last bullet point right. We've recently imaged an actual planet. And not just with the JWST. Using a technique called astrometry, scientists at the W. M. Keck Observatory on Maunakea, Hawaii, have also directly imaged a low-mass planet, AF Lep b, orbiting a young Sun-like star.

The hunt is on.

Life As We Know It

THE FIRST DISCOVERY OF LIFE BEYOND EARTH will likely not be when they visit us, or we visit them, but through something much more mundane: the analysis of the atmospheres of exoplanets.

A newly-named type of planet, called a Hycean world (blending the words hydrogen + ocean), is accessible to our most advanced instruments, and we may discover that these worlds support life.

Spectral scans by JWST of one of these Hycean worlds, K2-18b, have turned up enough information to begin piecing together an image of how this particular planet might be built, and we've discovered chemical signs that are associated almost exclusively with life. Additional scans over the coming year will determine if that signal is in fact cause for celebration.

Hycean worlds, defined by a worldwide ocean overlaid by a hydrogen-rich atmosphere, wouldn't be great places for us humans to live, but they *could* support life as we know it.

And that would be a major milestone.

Our Own Backyard

THERE ARE WORLDS RIGHT HERE IN OUR SOLAR SYSTEM that may contain life. Probably not intelligent life—at least not tool-making intelligent life—but life nonetheless.

How cool would that be?

Europa and Enceladus, moons of Jupiter and Saturn respectively, appear to have salty, liquid oceans beneath their thick, icy surface. These sub-surface oceans may have temperatures (due to internal heating) and conditions conducive to supporting life. It would be life quite different from what we're used to, but if discovered it would be one of humanity's major breakthroughs.

Missions to investigate have been approved, proposed or launched (the European Space Agency's JUICE mission; see below). Other moons in our solar system may have similar conditions, with a similar chance for life.

Did You Know?
As of this writing the European Space Agency (ESA) has launched a probe aboard an Ariane 5 rocket, which will arrive at the Jovian (Jupiter) system in July of 2031, with the purpose of exploring these potentially life-supporting moons.

The mission is called JUICE (Jupiter Icy Moons Explorer), and it will spend four years exploring Jupiter's Europa, Ganymede, and Callisto moons, looking for conditions that could host life.

Side Note:
This mission was the last liftoff for the Ariane 5, a true workhorse. What a perfect way to retire.

Follow the progress of the Ariane 6. It's needed quickly to replace the 5.

—

NASA has its own upcoming mission, Europa Clipper, the first mission ever specifically dedicated to studying that Jovian moon.

As well, Titan, the aforementioned moon of Saturn, has an atmosphere, with climate, rain, even liquid oceans. Features very similar to Earth. Those oceans are probably methane and ethane, but the right form of life might exist there as it does here. Could a silicon-based lifeform evolve in those conditions?

Future missions will find out.

Then there's Saturn and Jupiter themselves. As long as we're imagining unique forms of life, how about some sort of floating organism? Winged manta rays or the like, born and living their entire lives in the air?

Both Saturn and Jupiter would have a sweet spot at some altitude. A Goldilocks zone of their own (not unlike the 30-mile zone on Venus), where pressure and temperature are suitable for life to take hold.

What if there was a spark of life on one of those iconic gas giants, a billion years ago, leading to the evolution of an eternally flying or floating ecosystem? Much the way fish float in the ocean, those creatures would float in the air. Some sort of hitherto unimagined biology, leading to life as we barely know it.

Circling back for a moment to that intelligence thing, who's to say we might not find one of those species to be intelligent? Incredible air-dolphins, for example, running on some unique chemistry, swimming in that perfect zone in Jupiter's atmosphere, not a pair of pliers or a radio among them, every bit as smart as we are. Poets and music-makers, philosophers or telepaths, or any other form of intelligent expression that requires no machines, devices or technology.

Interesting concept, eh?

Out there?

Sure.

Impossible?

No.

With enough imagination it's easy to see the many possibilities that might exist right here in our own backyard.

DID YOU KNOW?

We've sent 48 missions to Mars.

We've been sending stuff to the Red Planet for some time now, and for the first time in human history we're taking serious steps to achieve the next milestone:

Putting one of us on its surface.

NASA has established the new "Moon to Mars Program" specifically for that purpose, and is now heading into a second phase that will focus on the Red Planet.

Things are getting real.

PUSHING THE ENVELOPE

Planetary exploration is not just for well-funded nation states anymore.

Private industry has been doing more toward the pursuit of space as the commercialization of the next frontier accelerates. As we've seen, space stations and other projects are already underway.

In one of the more recent such moves, Rocket Lab partnered with MIT to launch the first mission to study Venus. Purpose? To look for signs of life in the clouds.

Larger purpose? To demonstrate what these sorts of commercial missions are capable of.

Which is exactly what is needed.

These sorts of "grand reaches" by the private sector will not just continue to shove the doors to space wide open, profit-motivated industry and commerce will hold them open, once and for all.

Finally, there will be no turning back.

Space will become our new reality.

Abundance

LET'S TALK AGAIN BRIEFLY ABOUT SCARCITY. Specifically when it comes to resources.

When we talk about limited resources what we really mean is resources here on Earth. Saying we have a limited amount of gold on Earth is like saying we have a limited amount of Doritos in our house. A statement that's probably true, yet not the entire picture. A limited personal stash of Doritos in our house is likely—and perhaps quite sadly—the reality of our life, but if we have a car, a bike, or two good legs, there's plenty of stores in range to help us increase our stash. So ... when we say we have "limited" resources here on Earth it's really just a way of saying we need to figure out ways to get to the abundance of resources out there waiting.

Remember the Psyche asteroid?

As with everything, there's ultimately a profitable motive to learning all we can about our celestial neighbors. Asteroids, moons, planets and all.

As well as finding ways to easily get to them.

Adding to that prior section on *Space Mining*, "If we can find them we can mine them," might become a mantra for future resource-hunting space prospectors.

All the Doritos we could ever want are out there.

Let's think—and act—in terms of abundance.

It's the fastest way we win.

Pure Discovery

DISCOVERY CAN ACT AS A DRIVER OF PROGRESS as much as profit.

How epic was it when we took a photo of our first black hole? A monster in M87, a galaxy 55 million light years away. Back in 2019 we captured it on film, a beast 6.5 *billion* times the mass of our sun.

Three years later, in 2022, we advanced that same technique enough to image the black hole at the center of our own galaxy. Much smaller (4 million solar masses vs 6.5 billion), but also closer (27,000 light years), its mere existence was something we'd only speculated on not that long ago in our scientific development, now we've got pictures.

Pushing these observation technologies to the next level, and the next, and the next, advances our understanding of the universe around us.

Which, in turn, drives us to figure out ways to adapt that understanding for our gain.

In the coming years newer and even more custom-tailored and amazing telescopes are going up. One that will image more than 50 times as much sky as the Hubble (the Roman Space Telescope) is planned; another that could, among other things, allow us to spot planets in other galaxies, as well as detect extremely faint gravitational waves across the universe (LISA), is also in the queue; yet another that will scour a million stars looking for planets closer to home (PLATO). Massive phased arrays (groupings of huge antennas all over the planet, tied together by computers so they work as one) are coming online, being used for multiple observational purposes.

And on and on.

PUSHING THE ENVELOPE
Speaking of gravitational waves, being able to detect and measure these is a relatively new sensation. Info from such waves allows us to use gravity to explore the universe in the same way we've used electromagnetism until now. EM includes visible light to radio waves, all the way up to gamma rays, and is the way we've historically observed the cosmos.

That won't change, but being able to study gravitational waves promises to add a whole new dimension of discovery.

The new LISA detector, mentioned above, and other techniques being developed, will allow us to begin looking further back into the history of our universe than ever before. The early universe, prior to about 380,000 years after the Big Bang, was so hot and dense that it's thought to have been opaque to electromagnetic (EM) radiation. As a result, no standard telescope is able or will ever be able to peer any further back than that.

There would be nothing to "see".

However this is not a limit for gravity. Which means with the right detectors we can, using gravity waves, "look" all the way back to the beginning, giving us a first-ever glimpse of the universe during its remarkable inflation period. Being able to understand more of that era would be huge. Current theory holds that this inflation period was a brief time in which the universe increased rapidly in size, scaling up at a rate that, like many of these other cosmic concepts, is hard for our minds to conceive.

This was in the moments right after the Big Bang.

In essence, during that time there was a strange type of vacuum energy that caused the universe—the volume of space itself—to expand by a factor of 10^{78} in a fraction of a second.

Yeah.

Another stupidly big number.

That's an expansion of distance by a factor of at least 10^{26} in each of the three dimensions.

Can we conceive of that?

You know we love to try. Roughly, it would be like going from the size of a grapefruit to the size of a galaxy, in a second, or something equally incomprehensible.

Okay, so maybe we can't truly conceive it.

However, with this new crop of detectors, and techniques, we might get a glimpse into possibly the most epic early moments of our creation.

The Darks:

Another telescope, the European Space Agency's Euclid space telescope, just went up, on a mission to map the extragalactic sky in order to study the effects of dark energy and dark matter. If you haven't heard of the two 'darks', energy and matter, cosmologists suspect them but can't find them. Confirmation—and especially understanding—of either would mark radical new opportunities in physics, which consequently could bring about a shift in our technologies and what we're able to do.

One of these 'darks' might even shed some light (groan) on novel new ways to travel among the stars.

—

Throughout human history discoveries about the world around us have lead to new ways of looking at things, directing us in each case toward fresh focus for our goals.

Planets and cosmology are only the beginning.

By discovering the unknown we learn what's possible. And when we humans discover something is possible, it doesn't take us long to make it part of our everyday lives.

CHAPTER VIII
Physics

BRINGING US TO PHYSICS. The foundation of everything.

We hear your collective gasps:

Physics! Eep!

Some of you we're trying to make into space fans, and if you're not already very science-oriented a chapter like "Physics" just might send you running. Coming as it does on the heels of the semi-technical space stuff we've already been throwing at you, and preceding as it does the next chapter on chemistry (Chemistry! Double eep!), we realize for some this may be a bridge too far.

We might be at risk of losing you.

Why do we need to talk about physics?

Stick with us. Don't bounce. Physics is vital, as is chemistry, as you'll see, and in true "us" fashion we'll do our best to keep things engaging and fun.

We're Making Physics Phun!

Okay. Maybe that doesn't help.

But who knows? If we do our job right you might not only learn something, but you might just come through this in one piece.

Plus, after these two chapters, our story veers back toward the more generally interesting, with chapters on AI, robotics, the human condition and more.

Is that enough to keep you going?

Can we get you to read now and judge later?

Here we go.

From the moment everything came into existence (the Big Bang), physics has guided the formation of our universe.

Not surprising, then, that the study of physics can get rather complicated. Explaining all that and how everything else works takes quite a bit of math, and even more understanding.

It's also the key to our space future.

Physicists have been making discoveries throughout the years, building on the discoveries of those that went before, unveiling more and more about how our universe works. Other scientists and engineers are then able to take those discoveries and put them to practical use, inventing and building our latest technologies.

Some of the biggest names in physics, and their discoveries, are:

- Galileo Galilei (1564 -1642) – observations, experiments and mathematical analyses in astronomy and physics.
- Isaac Newton (1643 – 1727) – developed theories of gravitation and mechanics, plus he invented the math beloved by high school students everywhere, calculus. (We can hear the boos.)
- Michael Faraday (1791 – 1867) – discovered electromagnetic induction and devised the first electrical transformer.
- James Clerk Maxwell (1831 – 1879) – discovered radio waves, proving his own theory of electromagnetism.
- Max Planck (1858 – 1947) – formulated the quantum theory.
- Marie Curie (1867 – 1934) – made numerous pioneering contributions to the study of radioactive elements.
- Albert Einstein (1879 – 1955) – Einstein is best known for developing the theory of relativity, but he also made important contributions to the development of the theory of quantum mechanics.

- Niels Bohr (1885 – 1962) – made foundational contributions to understanding atomic structure and quantum theory.
- Richard Feynman (1918 – 1988) – known for work in quantum mechanics, the theory of quantum electrodynamics, as well as his work in particle physics.

And that's just the old guys (and only a small sampling). There are tons of new physicists doing all kinds of amazing new stuff, building on those earlier discoveries.

All hail your friendly neighborhood physicists.

Universal Constants

PHYSICS EXPLAINS THE FORCES, motion and energy of daily life. It helps us organize the universe. Physics helps us understand everything from exploding stars to the kinetic energy of a car crash, to the forces at work kicking a football, even the technology of our phones.

The laws of physics are universal constants.

Without an understanding of physics, from the basic to the advanced, we would never have been able to invent many of the things we take for granted today. Transportation, buildings, our electronics, new forms of healthcare ... basically everything we have and use in our modern society depends on this vital science.

That will hold true from here on out.

Yet ... those constants, whatever they might be, if we're doing our job right as scientists and explorers, will continue to be refined—and possibly even redefined—as we expand our quest for understanding.

In other words, constants exist, but we may have yet to fully realize them.

Science, lest we forget, is a fluid process of discovery.

Scientific knowledge is not static. It evolves.

A quote from theoretical physicist, Freeman Dyson, clarifies:

"The public has a distorted view of science, because children are taught in school that science is a collection of firmly established truths. In fact, science is not a collection of truths. It is a continuing exploration of mysteries."

Sure, it would be nice to have firmly established truths.

Truths *do* exist.

Annie sang that the sun would come up tomorrow, and it's good to have that particular truth to rely on. It helps us plan our day.

But the quest for knowledge means we never let our curiosity wane. We keep looking, keep studying, keep learning. Today we make effective use of the things we know now, even as we expand our effectiveness by making use of the things we learn next.

Always learning.

Always in search of the ultimate truth.

Things we were certain of a hundred years ago may have since been overturned. Things we're certain of now may be overturned up the line. Not all "known" facts are destined to fail, of course (the sun *will* come up tomorrow, after all), but it helps to remember scientific inquiry demands we continue asking questions and, most importantly, that we continue to seek answers.

The Big & The Small

AFTER THE TROVE OF AMAZING DISCOVERIES that have gotten us where we are in the world today, one of the most enduring, and quite probably most monumental, unresolved riddles in physics has to do with unifying the physics of the small (quantum) with the physics of the large (cosmological).

We have the math for the way big things behave figured out beautifully. It's called general relativity.

We have the math for the way tiny things behave figured out equally well. It's called quantum mechanics.

Or so it would seem.

However general relativity and quantum mechanics appear to be completely incompatible. The smooth, continuous universe general relativity describes conflicts with the discrete, chunky one of quantum physics. When you bring their equations together you get nonsense.

Therefore, finding calculations that work seamlessly to explain phenomena in both worlds will, it is believed, yield insight into many of the current challenges facing us. Do we have either of those disciplines (quantum or cosmological) wrong? Having such a unified theory could very well be the key that takes humanity to the next level.

Progress following such a discovery would then depend on us. Newton figured out orbital mechanics, but it was still a few years (well, centuries) before anyone orbited the Earth. Say the discovery of such a unified theory shows us clearly how to make wormholes or travel faster than light.

How quickly could we make those things happen?

If this generation of *Forty Suns* fans grows as rapidly as we hope, it shouldn't take long. The demand would be far too high.

DID YOU KNOW?

Random, cool thought for you. Speaking of the small and the large, did you know lifeforms on earth, humans in particular, sit pretty dang near the center of the universal size scale?

Crazy, but true.

A Planck length, named after Max Planck (one of our old dudes above), is the theoretical shortest length of anything possible.

How long?

A Planck length is on the order of 10^{-35} meters.

Admittedly, that's just another pretty meaningless number, and an even harder thing to imagine.

Rather than try to explain it, let us instead give you another unfathomable number.

Bear with us. This is going somewhere.

The observable universe stretches out on the order of 10^{27} meters.

Meaning ... wait for it ...

The entire observable universe is about as much bigger (more or less) than one meter as a Planck length is smaller.

Meaning things on the scale of meters (us, cats, elephants, etc.), sit pretty much at the middle of the size scale of existence. From the smallest to the largest, we're literally at the center of the universe.

What's our point?

Imagined another way, the smallest thing we know of is quarks, which are roughly 10^{-20} meters, which means if a quark were a person you'd pretty much be their universe.

Likewise you, to the universe, are a quark.

So the next time you're feeling small, especially in relation to the scale of our universe, remember you're sitting right at the sweet spot of existence. You are, in fact, a universe of your own.

Just try not to make any jokes about feely quarky.

Fancy Tools

FORTUNATELY FOR HUMANITY there are enough of us that agree on the importance of pushing our knowledge of these new frontiers. There are and have been and will be incredible people that know the value of these discoveries, and are willing to both fund and carry out the cutting-edge research that advances our knowledge.

To do that they make use of some pretty fricking amazing tools.

Heard of the Large Hadron Collider, for example?

This bad boy is the (current) king of particle accelerators, pushing the envelope, testing the very laws of physics, uncovering what the universe is made of and how it works.

It's an underground ring *17 miles long* near Geneva, made up of multiple detectors with multiple purposes, and it's been running since 2008.

How about the gravity detector, LIGO (Laser Interferometer Gravitational-Wave Observatory)? Another example of human ambition and ingenuity. Two ultra-precise laser detectors, each firing in arms 2 miles long, placed in entirely different parts of the country and working in concert to detect incredibly faint gravitational waves as they pass across the planet, allowing us to understand the workings of the universe.

Gimme some of that.

And we're beefing that up with the upcoming LISA detector from the European Space Agency (see the *Pure Discovery* section in the last chapter), which will use a system of space-going lasers and mirrors to create "arms" that are over a million miles long.

Then there's the Super-Kamiokande neutrino detector in Japan. Another engineering feat, the largest such detector, using 50,000 tons of pure water surrounded by 11,000 photomultiplier tubes buried 1 km underground.

These and so many other "fancy" tools help the physicists of the world make vital progress each and every day.

In fact, one of LIGO's lead architects, Rainer Weiss, had this to say when asked what the universe might still have in store for us:

"When we start looking with new and more sensitive instruments, better instruments, we turn up all sorts of surprises. That's happened over and over again in astronomy."

—

Planck got a lot of things named after him. One was a satellite, which was designed to measure temperature variations in the Cosmic Microwave Background (CMB) with unprecedented accuracy. The CMB sheds light on the very early universe, allowing scientists to understand the origins of galaxies and the condition of physics in the very first moments after the Big Bang.

The Planck observatory helped determine the age of the universe, the average density of ordinary matter and supposed dark matter in the Universe, and other important characteristics of the cosmos.

Fiction Becomes Reality

THE EVOLUTION OF OUR FICTION TO OUR REALITY is all but guaranteed. This has been the case throughout human history. We're too curious, too persistent for it to be any other way.

Think of just about anything, from airplanes to making fire, and it started as someone's imagination. An idea about how to do something that hadn't been done. The path to making airplanes and fire-on-demand—and every other new thing—involved increasing understanding of how things work, then taking that increased knowledge and applying it to practical use.

This is what we humans do.

It's what our space future is all about.

Bringing to mind another quote:

"There's always a wave front between our fiction and our reality. It's hard to know which fictions will become real and at what rate, but it's safe to say that with enough imagination all fictions eventually see the light of day.

"They are our dreams, and we have a habit of making them come true."

How about this for fiction becoming reality: Did you know teleportation has been done at a very small scale? Same with time shifting (restoring particles to their former state). Atomic-scale phases of matter have been found that challenge a fundamental law of nature: the prohibition against perpetual motion. Light-bending invisibility screens are now real. Others.

Though far from being useful yet, these sorts of discoveries are what lead to the paradigm-shifting ways in which we live our lives.

It only happens if we keep researching, keep building bigger and better tools—in short, if we never let our curiosity become fully satisfied. Dare to be bold, and commit to follow through.

Our tenacity is what defines us.

—

That last statement reminds us of a quick anecdote.

Humor us a moment. (As if you haven't been already.)

So far this lifetime your authors have been consummate jack-of-all-trades, which ... now that we're thinking about it, is probably why we don't have real careers and are writing this book.

Kidding.

Or ... maybe not.

We'll get back to you.

Moving on. One of the trades we held briefly was that of repo men. During our tenure a boss we worked for was so tenacious, so relentless in getting what he was after, we actually came up with a name for anyone with that level of assholish intensity:

Tenasshole.

Anyone that annoyingly dogged became a tenacious asshole. Not a jerk, per se, but so persistent it grated.

We're reminded of that because, though we humans as a whole aren't like that guy (trust us), in our own way we kind of *are* tenassholes.

We don't quit until it's done.

MOONSHOT MINDSET

Tractor beams. The stuff of science fiction but, like most of the cool things now part of our modern lives, something that could become science fact.

You might not know it, but tractor beams have actually existed in the real world for a while—at a very small scale. Microscopic tractor beams (also known as optical tweezers), are used to manipulate atoms and nanoparticles for use in medicine and research.

But they were always super tiny.

As of this writing there's been a breakthrough. We've now managed to make them big enough to see. A tractor beam strong enough to manipulate *macro*scopic objects, meaning you can watch them move with the naked eye.

How long before we're towing disabled starships with massive tractor beams?

—

Physics underlies everything we do.

In turn, the science and materials that construct our machines and our world are how we bring our ideas to life.

CHAPTER IX

Chemistry & Materials Science

BOTH CHEMISTRY AND PHYSICS deal with matter, the biggest difference being that physics is mainly concerned with how matter moves and interacts, while chemistry looks at the composition of matter down to the atomic level.

That's a simplification, of course, but for the speed at which we're plowing through topics here, it works. Forgive us, physicists and chemists out there.

Our hearts are in the right place.

Understanding materials is a key foundation of our technological advancement. From the materials that make up our batteries, to the materials and fuels we use to construct and power our rockets, how we build things determines how far we can go.

Strength

AS WITH MANY THINGS IN LIFE, the lighter and the stronger a material, the better. There are already so many great materials in existence today. We build modern airplanes by gluing together super-light composites. In fact, in terms of strength-to-weight ratios, the carbon fiber used to make planes is actually significantly stronger than steel.

But that same technology may not be best for all applications. Sometimes good old steel is the best solution.

What are some of the advanced things we're using now?

- Advanced Alloys (shape memory, high entropy)
- Advanced Polymers (electroactive, self-repairing)
- Biopolymers (DNA-based, protein-based)
- Porous Materials (microporous, macroporous)

Polymers include plastics. High entropy means high disorder, basically, with high-entropy alloys (HEAs) typically composed of elements with different crystal structures, giving them high strength, hardness, wear resistance, thermal stability and corrosion resistance.

Some classic fabrication materials:

- Tungsten
- Titanium
- Diamond
- Steel

Utility as well as strength plays a role. Gold isn't particularly strong, for example, but it makes a great conductor of electricity.

3D Printing

ONE OF THE LATEST MANUFACTURING TECHNIQUES is 3D printing. Also called "additive manufacturing", this is printing that, you guessed it, prints in three dimensions (3D).

It does this by using specific material compounds rather than ink. In this way the 3D printer is able to build, layer by layer, complex objects that are ready for everyday use. It started as a novelty, as many new technologies do, used to make models and figures from plastic, and has since evolved to machines that can print in a variety of materials, from titanium, stainless steel, advanced plastics, carbon fiber and many more.

A tiny few notable examples of crazy things that have been "printed":

- A titanium brake caliper
- A 2-story house
- A bridge
- A ship propeller
- A 25-foot boat
- An aerospace engine
- An actual *rocket* (see below)

And that's but a few stand-out examples.

Printing on demand will be a hallmark of future construction, and will allow us to not only build machines we need here on Earth, but on-site, at our future lunar and Mars bases as well.

Oh, and that rocket?

Pushing the Envelope
Relativity Space built and launched a 115 foot rocket, Terran 1, 85% of which was made of 3D printed parts. That's including the engines.

As of this writing they've now switched focus to an even larger version, the 270 foot Terran R.

Let's absorb that a moment.

A real rocket, a big one, that was launched, that was printed.

With more on the way.

Game Changers

Our amazing scientists of the world continue to imagine, test and invent new, advanced materials.

At this stage of humanity's technological evolution these new materials often have highly specific properties that outperform existing materials, enabling more innovative product designs.

Just a few of the materials that could soon transform manufacturing:

- Titanium Fluoride Phosphate
- Cellulose Nanofibers
- Self-Healing Gel
- Healing Metals
- Platinum-Gold Alloy (that just *sounds* expensive)
- Composite Metal Foams
- Spider Silk
- Shrilk
- Programmable Metafluids
- Carbon Concrete

Of those the one you're probably scratching your head the most over is "shrilk" (though spider silk is probably right up there; and honestly, the rest are more than worth a deeper dive when you have a few minutes).

Inspired by insect exoskeletons, "shrilk" is a biodegradable material made of chitosan, a component in shrimp shells, plus a silk protein called fibroin. It's as strong as aluminum and 50 percent lighter.

Then there's things like carbon nanotubes, lattice structures with an incredible tensile strength (resistance to stretching), in addition to other useful properties. Ounce-for-ounce carbon nanotubes are at least 117 times stronger than steel and 30 times stronger than Kevlar.

Ultra-hard starship hulls, anyone?

—

Certain materials grown in space are stronger and harder than those grown on the ground. Electronics that can withstand extreme operating conditions can be built using materials like silicon carbide, for example, which can be enhanced by being grown in orbit.

Yet another argument for larger on-orbit facilities and a greater commercial presence in space.

PUSHING THE ENVELOPE

At the beginning of the book we mentioned a definition for plasma, one of the four main states of matter (solid, liquid, gas, plasma). Never ones to sit still, at the bleeding edge of our science we've identified yet others.

States of matter called Bose Einstein Condensates—about as tricky and weird as they sound—are also in the mix. BECs are a state of matter in which atoms or subatomic particles coalesce into a single quantum mechanical entity (one that can be described by a wave function) on large (macroscopic) scales.

Such discoveries push our understanding of materials science, and thus what extraordinary applications we can find for the matter all around us.

In fact, just recently scientists from the US and China discovered a new state, called a "chiral bose-liquid", wherein electrons are essentially "frustrated" when trying to interact and find a place to "sit". Kind of like in a game of musical chairs.

We can relate.

It ends up having a useful result, in that electrons in this state will freeze into a predictable pattern. A stability that could have big applications for quantum-level digital storage systems.

Are you starting to get the idea we humans, right here on Earth, at this stage of our scientific and engineering evolution, are already pretty damn advanced?

In many ways we're already living our sci-fi future.

—

Room-temperature superconductors (materials that conduct electricity with no resistance), invisibility shields made of a novel material called broadband achromatic metalens, advances in graphene, quantum materials with exotic physical properties, plus many other inventions being imagined, developed or coming to market ...

The future is, truly, so bright, we gotta wear shades.

Preferably auto-polarizing ones made of advanced polymers with cutting-edge AR and VR (augmented and virtual reality) tech.

(For more on Alternate Reality see Chapter XII.)

Now on to more fun stuff.

CHAPTER X
Robotics & AI

As we put together this Zero Edition of *Forty Suns*, super jazzed about you and everyone else reading it, hopeful as all get-out that it will make even a bit of the impact we imagine it can, stuff is happening. Our tech world and, indeed, our world as a whole, is in the grip of what may well come to be known as the opening gun for the AI revolution.

We could be living at a moment in history that will one day be an entire chapter (possibly even a whole section of future libraries) in the world's forthcoming chronicles.

Yet it doesn't have to be a cautionary tale. Done right, this Pandora's box of possibilities, both good and bad, can be brought into use in such a way that humanity wins. This isn't the first time we've been down this road. If it seems a bit scary, remember hundreds of years ago the invention of printing spread ideas that played a part in causing devastating wars in the 16th and 17th centuries.

Eventually we got through it.

Now printing is a vital, foundational part of our civilization.

Not that we're suggesting AI will lead to wars; the point is, even such serious disruptions turned to a positive in the end. AI and the broad accessibility of the internet (social media) are comparable in their ability to spread disruption, possibly even chaos. But in the same way printing became part of the fabric of life, so too are these new technologies becoming.

It's simply up to us to make the best use of them.

Let's talk first, however, about robots.

Our Mechanical Companions

ROBOTS WILL GO PLACES WE CAN'T GO, do things we can't do. Whether directed by AI, by rudimentary programming or remote controlled by us, machines that take action on our behalf will become a key part of our space future.

Automated home vacuum cleaners are only part of the picture.

The current most common types of robot are:

- Autonomous mobile robots (AMRs)
- Automated guided vehicles (AGVs)
- Articulated robots
- Humanoids
- Cobots

Most of those are self-explanatory. Cobots not so much. Cobots are collaborative robots, meant for direct human-robot interaction within a shared space.

DID YOU KNOW?

As of this writing a trio of what are being called ANYmal robots, with fully-articulating, spider-like legs, built by researchers at ETH Zurich, are being tested in environments as close to the harsh lunar and Martian terrain as possible.

The team designed each of the three bots so they would be able to work both independently and together, the idea being that they will be specialized enough for particular tasks, but also similar enough to replace each other if one goes down.

Yet another cool idea being pioneered in robotics.

Power Up

INCREASINGLY WE'LL OFFLOAD dangerous and repetitive tasks to robotic machines. Surprisingly (or perhaps not so surprisingly) we've already built some pretty damn amazing ones. As with many of our technologies, the biggest hurdle is power. Fixed robots can be plugged in. If we expect them to follow us around the store, putting groceries in our basket, chatting up the cashier as we pay, they'll need their own power. We can make the robots; we have the tech to build powerful hydraulic and electric rams (the things that move the arms, etcetera); we've got the materials to make them strong, flexible, super-durable ...

We just can't run them for long.

When you fire up your robot companion in the morning it would be nice if it could hang with you all day. Go on your jog with you, toss the ball. Go to dinner. Politely laugh over a funny TV show, maybe even a guffaw or a knee-slap or two. Perhaps some *Dance Dance Revolution* before bed, a few video games. All without having to constantly recharge.

Everyone out there sweating over making our next-gen batteries, we salute you. The world is waiting.

Let us clarify. Those battery researchers and manufacturers *are* working hard toward each new advance. Demand, by which we mean profit potential, ensures the expansion of this frontier will continue to move fast.

The next-gen stuff will be solid state. Lighter, smaller, with potential run-times twice that of current batteries—plus charge times on the order of minutes or tens of minutes. Such solid state batteries are just over the horizon. Within just a few years we'll begin to see them in more and more applications.

With smaller better batteries our power demands come closer and closer to being realized.

Or even better, advanced power sources. Instead of better storage, let's make the power on-demand. A self-contained micro reactor that can be part of the robotic unit would be the perfect solution. Portable fusion packs?

Hells to the yeah.

We, your humble authors, aren't smart enough, but someone out there will be.

In the meantime our current crop of robots, and the ones being worked on right now, are still pretty cool.

Check out Tesla's Optimus, Figure's humanoid robot or Apptronik's Apollo, and anything by Boston Dynamics. (Speaking of *Dance Dance Revolution*, their bots have a few sweet routines.)

Did You Know?

Exo-suits are an actual thing. Also called powered exoskeletons, 'exo' meaning external or outside, these units are not strictly a robot, in that they're not autonomous, they're controlled by the wearer. However they provide robot-like capabilities, such as enhanced strength, greater endurance, and in every other way are like a robot.

The US Military is developing one. Dubbed the *Soldier Assistive Bionic Exosuit for Resupply*, or SABER (the military has the best acronyms), it's the product of a collaboration between the U.S. Army and Vanderbilt University. Slated to be deployed in the field the year this book goes to press, SABER will aid our men and women in uniform.

What's the most powerful exoskeleton so far? The Guardian XO, a robotic full-body suit manufactured by Sarcos Robotics (USA). According to Guinness World Records it allows the operator to lift objects weighing up to 90 kg (200 lb), with the operator bearing only around 5% of that load.

That means lifting your cousin Vinny would feel about like picking up 10 pounds.

Future astronauts will likely wear exosuits over their spacesuits, giving them strength-multipliers for moving stuff around on moon and Mars bases. With the lower lunar gravity, such powered suits would make wearers strong like ants. That 200 pounds would become a thousand (more) on the moon, with the same effort.

Will our future garages have one of these in the corner? Maybe one for each member of the family, next to the electric bikes?

Most likely. More than that, we're sure to see them in use as part of our exploits in space.

The 3 Laws

FOR FUN, we're including Asimov's oft-quoted Laws of Robotics. They make a whole lot of sense and may actually one day be implemented. (Bear in mind, these would more likely be Laws of AI, as a robot with no independent thinking could be programmed to behave exactly as desired.)

Here they are:

1. A robot may not injure a human being or, through inaction, allow a human being to come to harm.
2. A robot must obey orders given it by human beings, except where such orders would conflict with the First Law.
3. A robot must protect its own existence as long as such protection does not conflict with the First or Second Law.

That last one may be the stickiest wicket, to use a cricket term. Seems quite sensible, yet, ill-defined, we could end up with a robot (AI) with a skewed view of the importance of its own preservation. Interpretations as to conflict with the First or Second Law might lead to a decision preventing us from shutting it down.

These are fictional hypotheticals, of course, but we're getting pretty dang close to the point where we'll need to start giving them serious thought.

The Digital Mind

GOOGLE ANYTHING THESE DAYS RELATED TO AI and the top results are things like chess games, sex robots, etc. Not unexpected. We humans have our priorities.

However we all know AI will see way more use than that.

It already is.

AI today is like a really intelligent child. Genius but inexperienced. Think Frankenstein's monster, Anakin Skywalker (okay, maybe that's mean), Mister Data. Without the proper guidance, AI could get a little wonky. Experience is the beginning of wisdom.

In humans we often talk about IQ and EQ. IQ is, of course, Intelligence Quotient. How smart someone is compared to others. EQ is Emotional Quotient, or a person's empathy and ability to react properly and in the most pro-survival way to others and to their environment.

What is emotion good for and why does it matter?

Emotion is used to prioritize action. Like protecting someone, or opening a car door. Emotion motivates action.

What about for machines?

Machines don't have emotion, but might we develop a sort of AW, or Artificial Wisdom?

A fictional example of this would be C-3PO and R2-D2 from *Star Wars*. More human than machine, their emotion was just as likely to determine their actions as their machine intelligence.

Perhaps, as with human IQ and EQ, future digital minds might be rated according to AI and AW. Artificial Wisdom would instill reason or an ability to reason in our mechanical friends.

Types Of Artificial Intelligence

FOR NOW THERE ARE TWO BASIC TYPES OF AI, both starting with the letter "G":

- Generative
- General

Generative AI is what's making all the buzz right now. It's the "G" in ChatGPT, as in:
Generative Pre-trained Transformer (GPT).
As such generative AI isn't truly "intelligent". It just creates that illusion. Generative AI does this using what are called Large Language Models, or LLMs, which are, in essence, words and language data you, us, all our friends and everyone else around the world have been scattering by the crap-ton everywhere over the entire internet over the last many decades, which it pulls from and evaluates.
See? We used "crap-ton" again in a sentence.
And you thought we wouldn't.
LLMs are the result of applying major amounts of computing power to scour the Web and piece together massive quantities of unlabeled text, using self-supervised learning or semi-supervised learning to do so. (Side note, Nvidia, maker of the GPUs—Graphics Processing Units—used to do this, are rolling in dough thanks to the processing demands of AI.)
Generative AI is simply (if anything so amazing can be called simple) the result of wicked-smart instructions and programming, combined with next-level computational power, turned loose on reams of data. In essence, through its training, such an AI unit can parse inputs and, using the data in its LLMs, know what sorts of things it should "say" next.
Making it appear quite smart.
True AI, or "Artificial General Intelligence", is yet to come. This would be actual, problem-solving Artificial Intelligence; a form of

computational intelligence that would be able to replicate generalized human cognitive abilities. AGI, faced with an unfamiliar task, could find a solution. AGI, in its basic form, should be able to perform any task a human is capable of.

Fun stuff.

Pushing the Envelope

Each new version of ChatGPT pushes the envelope on the data used to form its base. GPT-3 used 20 billion parameters (a parameter is a value the learning system can change independently as it learns), GPT-3.5 uses 175 billion, and the latest, GPT-4, uses over a trillion.

Crazy amounts of data.

GPT-3.5 is primarily a text tool. GPT-4, on the other hand, is advanced enough to understand images. Provide it with a photo and it can describe what's in it, understand the context, and make suggestions based on that understanding.

As an example, people have used GPT-4 to craft recipe ideas based on pictures of the contents of their fridge.

—

The war in the field of Generative AI is exploding. Seeing what it can do (read: seeing the profit potential), the biggest tech juggernauts are rolling AI out as fast as possible, getting it into use everywhere—a mad competition that can only be expected to continue.

Most recently Google's DeepMind research lab unveiled plans for their own LLM, to one-up ChatGPT. DeepMind is using things learned from its AlphaGo AI system to make a ChatGPT chatbot called Gemini. AlphaGo, in case you missed the headlines, was the first AI system to defeat a professional human player at the board game Go.

Side bar, side note—whichever: The game Go is insanely complex. Makes chess look like checkers. Or, probably even worse than that. Tic-tac-toe or something. Despite its relatively simple rules, Go has roughly 2×10^{170} legal board positions. Yep. That's 10 followed by a hundred and seventy zeroes. Far greater than the number of atoms in the observable universe (estimated to be around 10^{80}).

What the what?

And so, yeah, not only was AlphaGo the first computer to beat the human world champion, it's now being used to found this incredible new Google AI system.

—

As noted, we're not up to the "general" AI stage yet. Until AGI becomes a reality, the current crop of generative AI offerings will continue to amaze, while finding use in pretty much every "in the box" application where there's money to be made. This includes creative writing, design, legal briefs, programming (especially programming) and many other tasks we, until only recently, thought to be the exclusive domain of real, live people.

That said, rest easy. The things that truly make us human are in no danger.

Yet.

Touched By AI

YOU'VE PROBABLY BEEN USING AI for some time without fully realizing it. Rudimentary forms of AI and machine learning, like predictive text, assistants like Siri, emoji suggestions based on message content and other examples have been with us for years. Not that we're rushing to normalize this amazing tech, but the truth is it *has* become normal, and future iterations will become normal as well.

AI will become part of our everyday lives.

It does, however, introduce considerations we've rarely had to face. The idea of "move fast and break things"—the way much human innovation to date has been achieved—may need a re-think when it comes to AI. Caution is advised. Prudence.

Unlike previous pushes into unknown fields and new tech, it may not be better, in this case, to act first and ask forgiveness later.

Else we might be asking forgiveness from the very machines we've created.

AI researchers have called it the "half-pipe of heaven and hell", meaning we tend to think of AI as either wonderful or terrible. Which isn't surprising. The singularity—when AI exceeds its creators, capable of improving itself and building technology more advanced than we ever could—probably isn't far off.

How will we remain in control past that eventual marker?

In the coming years it will be vital that we're exceedingly clever in how we roll out this civilization-changing technology. Not only in paying attention to its potential threats and misuses, but in seeing to its promise. AI is a different animal, and we want to pay close attention to how we regulate it.

Primarily because of the potential it holds.

Ideally we would avoid it being argued over in the political arena, as has happened with social media, for example.

Kent Walker, president of global affairs at Google and Alphabet, notes:

"Social media isn't going to cure cancer, but AI has the potential to, and it would be a shame if that promise were politicized. It would be a shame to hold back progress in nuclear fusion because we can't agree about Twitter."

AI will challenge us in many ways.

But if we navigate those waters successfully, if we find a way to be exceedingly smart about it, AI will reward us tenfold beyond those challenges.

Believe it or not AI has already peaked, for now, and in fact has actually waned when it comes to consumer interest—though due primarily to the passing fervor of the initial novelty-frenzy rather than a cooling of the technology. On the contrary, if anything AI is shaking down and firming up (sexy), becoming what will be as common a part of our daily lives as the internet or air.

AI & Space

AND WHAT ABOUT OUR SUBJECT MATTER, SPACE? Will AI help us get there faster? Will it invent an actual warp engine?

When it comes to space it will be interesting to see how these new computing resources are employed. Everything from mission planning to engine design may be impacted as AI advances and comes into its own.

Echoes of AI's potential can be found in Mr. Walker's quote above. AI may, indeed, become one of our best scientific R&D tools—if we let it.

Taking an elevated view of our history of innovation, there have been three distinct periods where we saw a paradigm shift in the way we do science. (Paradigm shift: a major change in the way something works or is accomplished.)

1. 17th century = microscopes and telescopes to observe beyond our own physical abilities; scientific journals to share and publicize discoveries; along with the wide use of the all-important scientific method.
2. Late 19th century = research laboratories, bringing together scientists and ideas in an industrial format.
3. Mid 20th century = computers, enabling new levels of simulation and modeling.

AI will likely represent the next paradigm shift, primarily in two ways:

- Analyzing existing scientific literature on a massive scale, searching for connections and drawing conclusions that may have been missed by its human creators among the vast troves of information. (*Perhaps a Roman Locomotive moment is waiting, if we unleash AI on the full body of our current scientific research.*)
- Using AI to form new hypotheses by analyzing existing data, then testing those results by posing and performing experiments much faster than human scientists could.

In the same ways each earlier shift brought with it major, positive advancements for the science of the day, AI may usher in a new era for our current development.

An Observation:

In some ways, handing humanity AI in the current age is like handing cannons to cavemen. Not properly understood the cannons could cause more harm than good. This much is quite true.

However that doesn't mean we should deny the cavemen cannons.

Cannons, properly employed, fundamentally understood, would be an important advance in mastering their world. Better survival and better conditions would follow. New technologies would evolve, much faster than otherwise. Such is the case with AI. We're at a similar stage of grasping the potential of this new tool, and in some ways it poses similar threats. Like the cavemen we lack a clear idea of the long-game repercussions if we use it wrong.

Like the cavemen, however, odds are very good we'll figure it out. And become more powerful in the bargain.

Pushing the Envelope

Human interaction isn't where AI ends.

AI is showing promise with other creatures as well. Digital bioacoustics is the study of the production, transmission, and reception of animal sounds, and it may be assisted by the same sort of data-crunching LLM technologies. Scientists are looking at ways to pair these technologies in order to both understand and communicate with animals.

In fact it's already been done with bees and bats.

Could this set precedent for future alien-language decoders?

It certainly won't hurt.

Our wonderful scientific minds (gotta love our scientists) are, even now, looking at ways to do this. In the same way AI will increasingly facilitate faster and better native-language translations, it might one day be applied for these other uses. Not only for better communication with the other inhabitants of our planet, from whales to our pet parakeets, it may prove useful in working out how to handle future encounters with other life.

Not The End, The Beginning

THIS IS THE BEGINNING OF THE AI AGE, not the end of humanity. We'll learn to cope, we'll figure our way through the challenges ahead and arrive at the best solutions. We always have, we always do, we always will.

It's said good judgment comes from experience. Unfortunately, most experience comes from bad judgment. A potentially catastrophic maxim where AI is concerned.

But we're hopeful.

Let's see how many mistakes it takes to get this right.

—

AI will, increasingly, handle the 'thinking' tasks we expend time on, freeing us for higher-level creativity. Future astronauts will be able to discuss situations with their AI helpers and fast-track solutions. We here on Earth will too.

Robots will go where we can't, do what we can't do, and that has huge value. Mechanical and computing assistance will be the backbone of our space future.

Our greatest achievements, however, will be the things we can experience ourselves.

CHAPTER XI
The Human Condition

As MENTIONED ALREADY, these human forms of ours are relatively weak. Yet, in some ways, they're incredibly durable. As living organisms we're resilient, dynamic. We self-repair—a thing that sets us apart from our machine friends. Machines don't wear out unless you use them. They don't get better unless you fix them. Bodies, on the other hand, require you to use them. To break them down in order for them to build back stronger.

When it comes to that cycle we tend to think only in musculoskeletal terms (how much ya bench?!), but as living organisms all our body systems operate this way. And we've got quite a few. Systems, that is. In case you haven't noticed, we're pretty complicated.

Each of us is walking around with a:

- Nervous system
- Digestive system
- Musculoskeletal system
- Circulatory system
- Endocrine system
- Immune system
- Respiratory system
- Integumentary system (our skin)
- Reproductive system

All get stronger through use.

(Let us pause to note that, yes, the jokes about the reproductive system getting stronger with use are not lost on us.)

The cycle is simple: We challenge our bodies, then we feed and rest them (well enough, hopefully), after which they, in turn, respond by toughening up for the next time they're challenged.

Call it the Tiger Cycle:

1. Chase/Kill
2. Eat
3. Sleep

You can fiddle that around any way you like, but the sequence is fundamental to the way our bodies work. We exert effort, we replenish, we rest. What's more, we can enhance those natural breakdown/rebuild cycles. Supplements help with muscle growth, varied diets challenge and strengthen our digestion, vaccines aid the immune system, stimulants improve mental acuity, exercises can be used to train our brains, etc.

How far can we take it?

The Outer Limits

THERE'S A BELL CURVE OF SORTS. A graph that defines the body's optimum range of use. With the muscles, for example, at one end of the curve you could sit on the couch and do nothing, get weak like a fat veal calf, get vulnerable and die. On the other end you could run continuously until you die. Somewhere in the middle is the sweet spot. Exercise is critical, but it needs to be done right. When it comes to the internal organs, the same is true. At one end of the curve, if you don't drink enough water you expire. At the other end, chug a few gallons in one sitting and you're probably also finished. Immune system, same thing. Wash your hands too much, over-sanitize, avoid all contact with dog kisses, germs and yucky dirt and the next big bug that comes along may just kill you. On the other hand, maybe washing your hands once in a while helps keep the balance?

There are many such examples.

The point is, with the proper awareness—and challenge—we can squeeze quite a bit of performance out of these otherwise weak pieces of meat.

What are the limits of the human body?

No one body can achieve them all. Each standout forfeits abilities in other areas, to a greater or lesser degree. The guy who deadlifts a thousand pounds probably can't run a 9-second hundred-meter dash. Nor can he, or the sprinter, execute a Front Swing Triple Lindy with Half Turn Dismount on the uneven bars. (We made those moves up, we think, but they sound pretty difficult.)

Form follows function; specialization is what gives us humans the advantage. One reason we can go further in any direction as a species is due to our allowing each other to specialize. If we all knew the same information, if we could all do more or less the same things, we'd be closer to ants.

Diversity is key.

Records give us an indication of what can be achieved. They put a marker post on the limits of the human form. Perhaps most importantly, they let us know just how awesome we can be when we apply ourselves to a goal.

A Few For The Books

TO GIVE AN IDEA, here are a few examples of what the human body can do. These are select areas of human achievement, checked at time of publication. (In other words, we humans, being prone to pushing ourselves like we do, may have beaten a few of these records by the time you read this.)

Enjoy.

Breath Holding

Get this. The record is 11:35. *Over eleven minutes.* Crazy? No doubt. Please don't try that at home. Or at all.

Bringing things a little more down to earth, Navy SEALs train to hold their breath underwater for two to three minutes or more. Breath-holding drills are typically used to condition and build confidence for high surf conditions at night.

We've probably all seen the 6 and 7 minute examples set by a few Hollywood movie stars while filming underwater scenes.

These achievements are a great example of how our physical limits are often beyond what we imagine.

Highest Vertical Leap
47.1"

For the average person, it's more like about 25 inches without training.

How high could that average person leap on other worlds?

- Venus   27 inches
- Mars   5 feet
- The moon   12 feet
- Titan   14 feet
- Pluto   30 feet

Deadlift

This one went into overdrive amongst lifters when Eddie Hall decided to round off to the nearest hundred. The record had been inching up incrementally, and was sitting at 465 kgs (1025 lbs), when, for giggles, Eddie threw on an even 500 kgs and ... actually lifted it. Not to be outdone, a few years later Hafthor Bjornsson did 501 (1104 lbs). Then, of course, someone else did one kg more—502—and now we're back to incremental gains.

But that's typical for competition. The coolest thing about this is that it's a perfect example of how we can blow past the bleeding edge when we decide—with conviction—to do so. Once someone sets their mind to doing a thing, then does it, the rest of us quickly follow.

All we need is that first pioneer.

Highest Human Speed
27.78 mph

(We're reminded of all those dramatic movie scenes with an action star leading a lengthy, harrowing chase down an alley, being pursued by a car. In truth that chase would never last more than seconds, even if the car started from a complete stand-still. Next time you're driving through your neighborhood, note how slow 20 mph actually is, and how quickly and unexpectedly you reach it.)

Sharpest Human Vision

This may not be one you can totally train and have control over, but it gives you an idea of what we might be capable of. An Aborigine man holds the record, clocking in with a 20/5.

What that means is that from 20 feet he can perceive details that an average person with 20/20 vision can only see from 5 feet away.

And yes, we know you're wondering: That measurement does in fact compare to the natural eyesight of eagles.

IQ

A score above 130 indicates exceptional intelligence. Anything higher than 160 makes you a genius.

Heard of Marilyn vos Savant?

Since 1986 she's written the column "Ask Marilyn" in the Sunday *Parade* magazine. According to Guinness World Records, her IQ of 228 is the highest ever recorded.

Hardest Punch

From brains to brawn. We don't know how hard Marilyn vos Savant punches, but it's probably not as hard as current record-holder Francis Ngannou. Heard of him? He's made a lot of news due to his MMA and boxing career, gaining more headline space than the punching king he dethroned, Tyrone Spong.

How hard?

Ngannou clocked in with a punch that, translated to energy, equals roughly 93 horsepower.

Yowza.

As you slowly wipe that shocked look from your face, realize that doesn't in any way mean Francis can power a small family sedan. There are far too many mitigating factors that go into that number. However, it does give you an appreciation for how much a smack from the MMA champ would hurt.

Reaction Time

This one hasn't been competitively measured, but the fastest conscious human reactions are around 0.15 seconds, with the fastest recorded times around 0.12 seconds (120 ms).

Most of us can achieve around a 0.2 seconds reaction time.

Unconscious, or reflex, actions are faster, around 0.08 seconds (80 ms), because the signal doesn't have to go via the brain.

Endurance

Ultra-marathons are probably the best example of human endurance. That these ultra-long run-fests are even possible, that they even happen is, frankly, remarkable.

It's basically anything over the standard 26 mile marathon.

Around 100 miles (160 km) is typically the longest course distance raced in under 24 hours, but there are also longer multi-day races of 200 miles (320 km) or more.

Memory

The longest sequence of objects memorized in one minute is 59.

The longest sequence memorized, period, is 100,000 digits over the course of 16 hours.

Highest Gees

Gees are used to express the force exerted through acceleration (stopping, going, turning). A "g" is equal to the force of one gravity, the pull you feel standing on the surface of the Earth.

Believe it or not, the highest recorded g-force ever survived is 360, by Karl Wendlinger when he crashed his F1 car at Monaco in 1994.

That's just hard to even imagine.

The highest sustained gees are, of course, fighter pilots, who experience the highest forces at up to 9 g.

PUSHING THE ENVELOPE

Outstanding physical and mental achievements have been inspiring us to better ourselves throughout human history.

We are, however, collapsing the frontiers of what may be physically possible for the human form.

It took quite a while to break the 4-minute mile, then we began breaking it routinely. Now even high school track stars can do it. Yet here we are, 70 years after that record was broken, and we've only managed to drop it by about 20 seconds, to 3 minutes 43 seconds.

That's our new standard, and it has been for nearly a quarter of a century.

In many ways we're up against the wall of physiology. Just how far can we push this bipedal organism of ours? How much was it made for?

Continuing in the running vein, the record for the 100-meter dash is still 9.58 seconds, set by the legendary Usain Bolt all the way back in 2009. Times since then have actually slipped.

What will it take to continue to advance what the human body is capable of? New genetics? Better diets in our youth? Programs to build better, faster, stronger bodies from a young age?

While it does appear we, as a species, may be nearing the limits of what we can achieve, there's no telling what the future might hold. Or what novel new ways we might come up with to exceed those limits.

Like most things, we have a habit of pushing the envelope until we can do more.

Ghosts In The Shell

ORGANIC BODIES ARE WHAT WE'VE GOT. It's who we are and, personally, your authors wouldn't want to change that. Being human is fun.

While we may not be built for space, science is forging ahead with all sorts of projects that will make it easier to live and work there. Enhancing the human condition is yet another area where our brightest minds have turned their focus.

In fact, let us pause.

To acknowledge the amazing players in the fields of medicine and biology working in these areas.

There are incredibly smart people, alive right now, today, working this very minute, heads down in their labs, engaged upon their research, explorers in their own right, blazing trails toward ways we can become better versions of ourselves.

If we humans are the future (and we are), then it is to these heroes of the human form, perhaps, that we owe the sharpest salute of all.

PUSHING THE ENVELOPE

Organics, in and of themselves, aren't necessarily a liability.

Take the incredible tardigrade.

Also called water bears, these tiny beasts are unique in their ability to survive in space. Tardigrades can live indefinitely in a vacuum, survive heavy doses of radiation, and withstand some of the harshest sunlight.

One incredible anecdote; there was some moss that had been dried out that had been in a museum for over 100 years, and when they moistened it ... out crawled tardigrades!

So many cool bits of research on these little dudes.

They're tiny, about the size of one of the periods in this text, and with the right light can be seen with the naked eye.

Long story short, these are a living example of the degree of toughness organics can conceivably achieve.

Hm. Maybe we should become tardigrades.

Cool as that might be (or maybe not), our two main (realistic) options at this time, when it comes to these shells of ours, are to beef up the organics, or add artificial enhancement.

Bionics

WE HUMANS AREN'T ROBOTS. If we were we could pick out all sorts of modifications for ourselves, from colors to styles, shapes to sizes, functions to capabilities.

Though limited in what we can do, there are still ways to mechanically enhance what we have. Bionics, a portmanteau (a blend of words) of the words "biology" and "electronics", is the replacement or enhancement of organs or other body parts with mechanical versions.

The bionics industry has grown along four major areas of application:

1. Vision
2. Hearing
3. Orthopedics (musculoskeletal)
4. Implants

Implants to date have been designed to augment cardiac and neurological functions.

Note, things like powered exoskeletons, while cool, are more something to be worn than integrated with the existing body.

(If you skipped the last chapter, Chapter X, see the "Did You Know" section on exo-suits. Another example of super-cool sci-fi tech coming to a reality near you.)

Cybernetics & Cyborgs

AS DISTINCT FROM BIONICS, cybernetics is typically concerned with automatic control systems, such as the nervous system and brain, or/and mechanical-electrical communication systems, and includes a wide range of systems that aren't always human dominated. In other words a "cybernetic" system could have as its prime director a human brain or something else.

Meaning the line can blur, as bionics (above) controlled by interfaces with the user's brain bleed into the cybernetic arena.

A "cyborg"—another portmanteau, this one of cybernetic and organism—would be a being composed of both organic and artificial systems. The key there being "a being". A cyborg, in keeping with that cybernetic mandate, could also have as its prime director a human brain, or something artificial.

The thing that makes it a cyborg is that it has organics and mechanical stuff.

PUSHING THE ENVELOPE

Brain-computer interfaces (BCIs) are a direct communication pathway between the brain's electrical activity and an external device, most commonly a computer or robotic limb.

These have been developed and are actually in use. In fact they're being refined every day. Three main types exist, intended for use where the BCI is either:

- Non-invasive (usually tapping externally recorded signals, like an EEG)
- Semi-invasive, or
- Surgically implanted.

These invasive implants are sometimes called BMIs (Brain Machine Interfaces). There's even a version being worked on that would bypass the nervous system in the spine to help with spinal injuries.

With the most effective BCIs and BMIs, we'll be able to control complex machines with our mind.

Enhanced Organics

ENHANCING THE BODY NATURALLY, or at least without mechanical aid or components, is another way to get more performance out of our future space-living selves.

Surgery might become an option, with the ability to somehow bolster those various systems for extreme environments. The broad list of ways we could enhance a human body are:

- Drugs
- Hormones
- Implants
- Genetic Engineering
- Surgeries

All this reminds us of pen-and-paper RPGs (roleplaying games) we played in the 80s and 90s. In those with post-apocalyptic settings (our faves were *Gamma World* and *Rifts*) there were character classes called Juicers. Rather than cybernetics, a player could choose to augment their character with a ridiculous pharmacopoeia of enhancement drugs. Strength, speed, alertness,

IQ—everything could be throttled up, using purely chemical and organic options.

Think Bane from Batman.

The theory behind it is sound; the body is a chemical system, after all, so adjusting that chemistry should yield results.

Such a future would be another case of fiction becoming reality.

—

When it comes to space, many of our tissues don't fare well in microgravity or micro pressures. In recent experiments where twins were compared, the one that spent extended time in orbit had tissue separation in the eyeball and other places that the Earthbound one didn't.

These and many other challenges for the human body will need to be addressed, with solutions developed.

Feeding The Machine:

Eating is something pretty much everyone loves to do. We have, like, a dozen TV channels dedicated to food, from meal prep to baking, competitions to restaurant overhauls, exotic cuisine to nutritional tips and how-to's.

We love it.

Maybe because we all need to do it?

Kind of like breathing, feeding these organic bodies is pretty important to keep them making energy and staying alive. Mostly we enjoy this process, when we can—which is probably why we have so many shows dedicated to whetting our visual appetite and stimulating the desire to do more eating.

What about in space?

NASA, probably the biggest expert on eating in space, describes it like going on a long-term camping trip with your closest friends. Storage and disposal are important. Preparation varies with the food type. Some foods can be eaten in their natural forms, such as brownies and fruit. Other foods require adding water, such as mac-and-cheese or spaghetti.

Of course considerations must be made. For example, salt and pepper come in liquid form, because astronauts can't sprinkle salt and pepper on their food in space. The salt and pepper would simply float away. Not only that, there's a danger the grains could clog air vents, contaminate equipment or get stuck in an astronaut's eyes, mouth or nose.

Nutrition is the centerpiece of any of it.

Call it "tactical nutrition". Eating for the purpose of optimally fueling the body. When it comes to mission readiness the nutritional value of the food is as important as other factors, if not more so.

Still, treats need to be factored in. We are human, after all.

Who knows what future developments will bring?

Could you be one to come up with a way to bake a perfect birthday cake in space?

Did You Know?
To get us ready for Mars—and other such longterm missions—scientists have conducted and are conducting fun (fun for the rest of us, anyway) experiments on willing volunteers.

We mentioned earlier the CHAPEA (Crew Health Performance Exploration Analog) experiment being conducted by NASA. Another, Mars-500, was a joint operation between the European Space Agency, Russia, and China from 2007 through 2011.

Mental (and other) breakdowns on an actual Mars mission could cause a very expensive operation to fail, which is why these studies are so crucial. They give us a basic understanding of how humans can function both getting to and living on our closest planetary neighbor.

Though a year-long isolation study is an intense way to test the limits of humanity, it's the best way to prepare for the future colonization of our solar system.

MOONSHOT MINDSET

How about time spent in transit?

Remember the hibernation pods in the movie *Alien?*

Experts believe we could be testing such long-flight hibernation within the next 10 years. Hibernating astronauts on a year-long trip to Mars would not need to eat or drink, and would require far less oxygen. Cost reduction, plus no need to keep occupied during those 12 months? Count us in.

There are other benefits.

Bodies of hibernating astronauts might waste away much less than the bodies of those awake in microgravity. Upon arrival, these hibernators would thus be fit and ready to commence challenging activities and exploration almost straight away.

It's called a torpor state, and it's like a pause button for the physical processes we'd want to put on hold during a long trip.

Additionally, the slowed-down cells of a hibernating body might not be damaged by radiation—one of the biggest health concerns of long-duration space missions.

Radiation

SPEAKING OF THE "R" WORD, as far as challenges go radiation will be one of our biggest, if not *the* biggest. Even as we advance all the other tech that will be needed to launch lots of things far and away into the solar system, to get them here and there super fast with minimal fuss, making space travel a snap ...

True colonization and long-term exposure will face this last, quite significant hurdle.

As such, we're giving radiation its own short section.

You may have heard of rems and rads when it comes to measuring death rays; the standard we go by these days is the sievert, which is a measure of the amount of radiation absorbed by the body. It's usually stated in millisieverts, or mSv, and NASA has an upper limit on how much their astronauts are allowed to be exposed to over the course of their career.

After that they're no longer allowed into space.

It's that critical.

Radiation messes big time with our DNA. We're not designed for the harsh reality of space. (Did we say that already?) We like it right here inside the protective bubble of Mother Earth's embrace, where her super-jacked magnetosphere blocks most of the harmful solar and cosmic rays that would otherwise ruin us.

Like all other major obstacles, therefore, radiation is a biggie we'll need to solve. But, and again, there's this one thing we humans do very well:

Solve challenges.

We're pretty dang great at it, actually.

One promising possibility, one that's already been tested quite a bit, is gene replacement. If we were able to simply replace damaged genes as they're broken we could reduce, if not eliminate, the effects of otherwise deadly radiation.

Heard of CRISPR? Clustered Regularly InterSpaced Palindromic Repeats?

"Ah! Yes," we hear you nodding thoughtfully, as you scratch your chin. "Clustered Regularly Inter ... yes, of course. That thing."

CRISPR is a gene-editing technology that made lots of headlines a few years ago. In short, it makes it possible to correct errors in the genome and turn on or off genes in cells and organisms quickly, cheaply and with relative ease. Which could include rapid repair of radiation damage.

(A palindrome, by the way, is a word, number, phrase, or other sequence of symbols that reads the same backwards as forwards.)

Shielding on our ships and our bases and our space suits helps, of course, and will no doubt always be used, but what if we could simply "fix" ourselves on the fly, undoing damage as it occurs?

Such a feat would quickly eliminate that one, huge barrier to a life in space.

Our True Measure?

CALL IT A BRAIN, call it a mind, call it our consciousness, our personality—we're talking here about the thing that truly is *us*. Who *we* are. Not how tall or how old, but *who* we are.

The real us.

Bodies define us; our spirit is what moves us.

Sure, the guy who can beat everyone arm wrestling might also be the guy who builds a civilization, but the fulfillment of that amazing dream doesn't happen simply because he has a killer grip. Humanity's greatest achievements come from the application of thought, imagination, intelligence. The real us. Which in turn directs the force that gets things done.

Can we measure that fundamental "cause"? The "prime mover unmoved", to give it a name? Our conscious awareness? The thing from which all else springs? The genesis of all creation; the "me" in each of us that dreams and then makes the world?

IQ is the way we measure intelligence. Intelligence may or may not be the best way to codify that intangible force of inception we each represent, but it works well enough.

It's all we have at the moment.

So, let's take a look.

Actually, let's stretch those mighty minds of ours and imagine what an even broader scale of IQ might look like.

Are the numbers we use to measure IQ all there is?

Is Marilyn vos Savant truly the pinnacle of what's possible?

An IQ of 228? The world record?

What would an intelligence beyond those ranges look like?

What if the IQ band is actually far wider?

Could it be?

As a comparison, in the electromagnetic (EM) spectrum visible light occupies a small sliver of possible wavelengths. For most of our human existence that narrow range was all we were aware of. It was all we could perceive. No one knew of microwaves or infrared. Couldn't feel them, couldn't sense them. X-rays? Didn't even have a name for that. A complete unknown.

Now, of course, we see just how much EM bandwidth exists. Ranges we need special instruments to measure. You may not be able to perceive the radio waves coming into your car, but with a "radio" you can hear them.

What if IQ has a similar phenomenon? What if our current measure of intelligence is something like the visible light band? With ranges beyond our current perception? What if there are IQ "wavelengths" far above anything we've yet achieved?

Could there be a level of intelligence that makes Einstein, or even Marilyn vos Savant, look like a slug?

Zero is zero, let's assume, so the IQ scale probably wouldn't go down. Even in harder sciences, like Physics, absolute bottom-scale stuff exists. Absolute Zero as a temperature, for example, is a real thing. The state where nothing is moving.

There's a theoretical upper limit too, but it's way, way up there. You think the core of a star is hot? That's an ice cube compared to the highest temperature theoretically possible.

Okay. Let's diverge briefly, because we hear you wondering.

That highest theoretical temperature is called the Planck temp (there's that Planck guy again), the theoretical upper limit for the measure of how hot stuff is, and it's 100 million million million million million degrees.

Nothing should ever be able to be hotter than that. The reasons for that aren't important here, and honestly the information itself probably isn't either, but we know you're curious like us so there you go.

And yes, that looks like we just got silly and typed "million" a bunch of times, but that's actually the right number of millions.

Pretty dang hot.

Back to the topic at hand. What if our IQ scale is similar?

Could there be ultraviolet-scale IQ numbers?

Gamma-ray mega-geniuses?

That sort of conjecture at this stage of our development leads to nothing particularly useful, but it's a fascinating thought exercise nonetheless.

What *is* our true measure?

When it comes to thought, creativity, understanding ... how much might we truly be capable of?

How much can we achieve?

How high can we go?

Us Among The Stars

ULTIMATELY, OUR OBJECTIVE IS TO BE OUT THERE, living among the stars. How do we become more effective spacemen?

Humanity has reached a stage where, within a few generations, we'll outgrow our home. To use an organic analogy (since this chapter is loosely about organics), we either stay in one place, one size, different flowers blooming in cycles but no real change, possibly even withering in place, or we spread beyond our single pot. We go further into the world (space), evolving as we expand.

We're big enough, and capable enough, to do that now. To spread our seeds, making more and more of us, putting down roots in more and more places.

Do we want to stay still? Be a lone, perfectly manicured bonsai?

Or do we want to become a forest?

Why not both? A beautiful home and a forest of other options out there.

For the first time in our history we can make that leap.

In many ways we already have.

Humanity is already spacefaring. The spacefaring part is really, really small compared to the whole, but we're doing it.

People are living in space.

Is it enough?

Of what can we be certain?

One thing's for sure: It's certain that one day we'll live among the stars.

That much is inevitable.

What's less certain is when.

Time is the only true uncertainty when it comes to our interstellar future.

We may first have to get hit by giant rocks. Or maybe we'd rather wait for a few super volcanoes, earthquakes—pick your Earth-indigestion event. Bad weather, climate catastrophe, slow collapse. Or some kind of crazy solar eruption. Or a more likely scenario, we decide that, before we truly get the show on the road, before we work together toward space in a big way, we first need a few more fights. You know, for old times sake. Really big ones, even, where we smack ourselves into apocalyptic oblivion—perhaps again and again (why do it only once?)—and only then, after we finally get it out of our system, do we make the leap.

After all, there will always be a few of us left. We'll rebuild, thousands of years will go by, maybe more, we'll screw it up again, but we'll survive, again and again, and, one day, one eventual day, we'll decide enough is enough and we'll get out there.

No biggie, right?

How many setbacks before we get it right?

A better question:

Why waste all that time?

Why not do it now?

There are very few good reasons why we can't.

Do we want to be the ones in future history book chapters covering the sad tales of the people that failed? Those poor fools who couldn't get out of their own way?

The alternative is to get out there, now, en masse, as fast as possible.

Whether we're personally the ones to go or not, we, as a species, should at least be *able* to be out there. Much like in the same way we ourselves may never go to Fiji, people should be able to go to Fiji. If someone *wants* to go to Fiji … that should be an option.

Agreed?

—

How we humans do that (go to space, not Fiji) goes beyond simply getting there. Much will need to be done between now and then to prepare us for such a life, to increase our resilience to radiation and other factors, to toughen us against the things that will threaten us every minute of every day we're away from the natural environments that nurture and sustain.

As long as we're human, the 'human condition' will be the thing we most urgently need to address.

Earth is our home, this is where we thrive, these are the conditions we've evolved to expect, and unless we find other places a lot like it, we'll either have to adapt to those new places, adapt them to *us*, or live within environments we create.

Thankfully entire industries are focused on solving this.

Our ability to figure it out has never been in question.

One day we *will* live among the stars.

CHAPTER XII
Communications & IT

OUR INTERCONNECTED WORLD OF computers and networking, both local and global, how we exchange, process and pass information is, truly, the foundation upon which all of this is built.

What kind of advances can—and should—we expect from the communication and information technology (IT) systems that underpin our world?

Home Grown

RIGHT HERE ON EARTH IS WHERE IT ALL BEGAN. Communicating ideas, coordinating action at a distance, relaying information. Language, the ability to synchronize activities—including the communication systems to do it with—has been the backbone of our human expansion. Increasing that capacity to collaborate and coordinate is what's allowed us to become the global civilization we are today.

A steady progression of ranged communication systems has, therefore, been vital.

Shouting was probably first.

Hand signals.

Runners would've been next, delivering messages. Signal fires. Mailed letters. Later, sailors used semaphores and flags, maybe the occasional flashed lights. Being human and needing to communicate more, we eventually figured out faster/better ways. Hard-wired communication networks, telegraphs, came into wide use during the late 1840s. Usually tracking railroad lines, transmitting information quickly over great distances.

The first trans-Atlantic line was laid in 1866.

After that came wireless. The first radio broadcast happened on November 2nd, 1920. (Though all the way back in 1895 a young inventor named Gugliemo Marconi was experimenting with using radio waves to transmit Morse code.)

From there we moved to phones, faxes and, of course, the internet. The first workable prototype of our beloved internet came in the late 1960s, with the creation of ARPANET, the Advanced Research Projects Agency Network.

January 1st, 1983, however, is considered to be its official birthday. Before that the various computer networks didn't have a standard way to talk with each other. After that date we had a standardized comm protocol—basically, a transmission language all computers spoke—and we were, as a culture, off to the races.

These days we can't live without it.

A few fun factoids:

- There are 15 billion internet-connected mobile devices in the world at present (though there are only about 8 billion of us walking around).
- Data sent each day among those devices? About 2.5 million terabytes.
- The amount of data currently on the internet is equally hard to fathom. A billion is big (a billion seconds ago it was 1991). A trillion is a thousand times bigger. Already there's no good way for the human mind to reference those numbers. However, for the hell of it, count to a trillion, then do that again, a billion times (count to a trillion a billion times). That's what's called a "zetabyte". As of 2023 the internet housed 120 of those. In other words, a lot of frickin cat videos.

Clearly communication here on Earth has been heating up. Big time.

No duh.

When it comes to space good-old radio comms are still the primary way we connect, using ground-based antennas and relays. Though that's improving. Add to that tons of data links here on the ground via giant undersea cables, more antennas, kicking things back and forth between themselves and through satellites in orbit, cell towers streaming furiously, wi-fi, fiber-optic cables and old, legacy copper, plus the new crop of low-latency, low-orbit satellite "constellations" designed to blanket the Earth's surface with internet access (like Starlink, for example), along with all other nodes and methods of access around the world that have been added and added some more, enhanced and refined, such that we are now connected like never before, and you're only just starting to get a picture of how far we've come.

These days it truly is difficult to be out of touch with anyone, or any bit of information.

—

Interesting tidbit. The sun blasts radiation into space across the entire electromagnetic (EM) spectrum, including the narrow bands where our electronic devices broadcast. We now spew out more in those fractional spectrums than our own sun, such that if you were to look at our solar system from far away, and only through those very narrow bands where we transmit, our Earth would appear brighter than its own star.

Welcome to the 21st Century.

DID YOU KNOW?
The NASA Deep Space Network (DSN) has been operating since 1963 (over 60 years!) and is a worldwide network of American spacecraft communication facilities, located in the United States,

Spain, and Australia. The DSN supports NASA's interplanetary spacecraft missions.

"Deep Space" in this case is anywhere from near-Earth, all the way out to the farthest planets of the solar system.

Cool fact: Due to the DSN's geometry, once passing an altitude of 30,000 km (19,000 mi), a spacecraft is always in the field of view of at least one of the tracking stations. Meaning if you ever find yourself further away from the ground than about 20 thousand miles—and you have the right gear—you should be able to reach someone.

Orbital

ON THAT NOTE, LET'S TAKE IT INTO SPACE.

That is, after all, our goal.

Data connections in space will be vital, relays for both intra- and inter-station communication, comms between satellites and other spacecraft, including what we'll use on future lunar and Mars bases, within the bases themselves, their compounds, remote vehicles and orbital outposts.

In many cases existing network technologies will do the trick.

New stuff, however, is also being pioneered.

Systems for broadband wi-fi in space-to-space applications are in development, better systems for station-to-station comms, etc., improving the speed at which data can be transferred (comparable to the speed of broadband we're used to on Earth), and so on.

Our communications needs in space will be as crucial as they are here at home.

Perhaps even more so.

DID YOU KNOW?

Higher bandwidth is coming to a space mission near you. Back in the day Apollo radios sent grainy black and white video from the Moon. With higher data rates, via an upcoming optical terminal on the Artemis II mission, we'll get 4K, ultra-high-def video beamed back to us direct from lunar orbit.

It will almost be like being there.

Pushing The Envelope

An upcoming mission to orbit, called *Polaris Dawn*, will, among other things, test advanced communications which will likely see wide use in future space applications.

From the mission plan:

"The Polaris Dawn crew will be the first crew to test Starlink laser-based communications in space, providing valuable data for future space communications systems necessary for missions to the Moon, Mars and beyond."

Note, this is the first of three planned missions in the Polaris program, and is worth taking a moment to look into. Founded and operated by a who's-who of power players in the space industry, not only is this program breaking new ground on several fronts, it supports a noble cause.

Latency

LATENCY (THE DELAY BETWEEN TRANSMISSIONS) gets higher and higher the further apart the terminals. No matter the extreme amount of bandwidth, transmission speed is constant and depends on distance. On Earth, or nearby, this delay is negligible. In the case of a lunar transmission (talking with people either in orbit around the moon, or at bases on the moon), the delay is small but noticeable.

These delays will need to be taken into consideration as we push further afield.

A few of what will likely be our most common transmit times:

- Moon = 2.5 seconds
- Mars (closest approach) = 4 minutes
- Mars (farthest) = 24 minutes

Communicating with our deepest solar system probes can take up to 4 hours or more.

Did You Know?
Interestingly, older, yet super-reliable tech, like Ham radios, can reach the ISS. Ham radio is still around and in wide use among enthusiasts. (Named after the first station call in 1908, made by the Harvard Radio Club, Ham Radio is amateur radio.) More powerful than Citizen's Band (CB), with a cap of 1500 watts transmitting, versus 4 watts for CBs, ham radios can reach quite far.

People use ham radio to talk across town, around the world, or even into space, all without the Internet or cell phones.

If you're looking for a cool hobby, with lots of history—that also ties into our space program—this could be your calling.

Get it? Calling?

Trust us, we need very little encouragement.

Pushing The Envelope
Synthetic Aperture Radar (SAR), is a new satellite technology which can penetrate most any atmospheric conditions. Lots of great applications for this. But it takes transmitting a *lot* of data. NASA and other organizations, like SpaceX, have been experimenting with—and making breakthroughs on—developing new protocols to encode this info in optical wavelengths and transmit it through lasers in space.

Space lasers!

There's that Dr. Evil image again.

Positioning those lasers, however, so their signals are received on the ground, especially from satellites moving at a speed of 17,000 mph in low-Earth orbit, ain't easy. Yet the promise of, and the demand for, this technology is such that the smart people that make these sorts of things work are, well, making them work.

By the way, that optical terminal we mentioned for the Artemis II mission? The one that promises 4K, ultra-high-def video beamed back to us direct from lunar orbit?

This is what it will use.

Networking

Mesh networks will likely be a standard for data transmission as we establish our presence in farflung locations throughout the solar system.

The official definition:

"A mesh network is a network topology in which the infrastructure nodes connect directly, dynamically and non-hierarchically to as many other nodes as possible and cooperate with one another to efficiently route data to and from clients."

Non-hierarchical means there aren't different levels of importance or status, which means each device, or node, can find the best solution for transmission.

At present there's no GPS or Wi-Fi around or on the moon, which means missions rely on constant supervision from engineers on Earth. As our presence there increases that will quickly become unsustainable. Hundreds of missions are planned to be launched over the next decade.

To fill that void, NASA has awarded a contract to Nokia to deliver internet to the moon. They're aiming to send the first 4G network to the moon by 2024.

PUSHING THE ENVELOPE

Reinforcing the notion of speed and data volume, the TeraByte InfraRed Delivery experiment (TBIRD—another great acronym), from NASA and MIT, which uses cutting-edge optical technology, recently broke the record for fastest laser communication from space—a record it held—at 200 gigabits per second.

Take that, radio waves.

Comms tech is on fire.

More Power!

BEHIND ALL THAT COMMUNICATION LIE OUR DEVICES. Our phones, our laptops, our larger computers, data centers and all else are the endpoints for those fantastic networks.

For the longest time we measured computing progress and power by something called Moore's Law. That was a concept put forth by a guy named Moore (you probably figured that out) back in 1965, when he observed that the number of transistors in a dense integrated circuit (our computer chips) doubled about every two years.

After a while it turned out to be an uncanny prediction, and our computing power did, indeed, double about every two years, for a really long time. In fact that geometric progression has only just begun to taper in the most recent decade.

Moore never knew how that simple observation would come to influence long-term planning and the setting of R&D (Research & Development) targets across the industry.

The main reason it's slowing down is we're now reaching size limits for transistors. Gates, the small structures that hold and direct charge—the fundamental components of our chip architectures—have gotten so small, so thin, electrical charges can jump across when they're not supposed to. As transistors approach the size of a single atom, their functionality begins to get compromised due to the behavior of electrons at that scale.

Of course, though Moore's law is now less and less applicable, our demand for more power, more speed, will never abate.

We humans have a habit of demanding More Better Faster.

Always we want more. (God bless us).

And so we innovate.

Things like three-dimensional (stacked) chip architectures, new materials that allow us to make smaller and smaller gates (transistors), biological and semi-biological computers made from living cells that make use of neuron-like structures to perform computations, using light instead of electricity (optical/photonic transistors), different types of specialized chip designs, multiple chips bent toward a single task, such as with super computers or grid computing, techniques like edge computing (processing data close to where it lives, enabling faster processing), plus a myriad of other ways in which we squeeze maximum performance from our devices to meet our needs will continue to push the envelope of our computing speed and power.

PUSHING THE ENVELOPE

Heard of "magnonics"?

Think electronics (electrons/electricity) or photonics (photons/light), but with magnetism.

Magnonics combines the study of waves and magnetism, with the aim of applying the behavior of spin waves to practical use in nano-structure (super small) elements.

Spectacular.

Dirk Grundler, head of the Lab of Nanoscale Magnetic Materials and Magnonics (LMGN), who's working on this, had the following to say:

"With the advent of AI, the use of computing technology has increased so much that energy consumption threatens its development. A major issue is traditional computing architecture, which separates processors and memory. The signal conversions involved in moving data between different components slow down computation and waste energy."

Magnons may be a solution. They can be used to encode and transport data without electron flows, which involve energy loss through heating. Magnons wouldn't suffer that loss.

It's a new tech which has the potential to reduce huge inefficiencies, thus providing a giant leap forward in supporting the demands of big data.

Thus helping to keep us on track with that Moore's Law thing.

Our Quantum Future

ONE RELATIVE NEWCOMER that's been gaining headlines is the concept of a quantum computer. First proposed way back in 1980 by one of our Physics giants, Richard Feynman, operational quantum computers will one day find use in many areas. Not only will they bring new solutions—and headaches—to security and crypto, perhaps one of the biggest things they'll do when fully evolved (and the thing of greatest interest to Feynman) will be to simulate conditions in our physical universe right on a chip.

Making them a huge tool for science.

Most quantum computer headlines tout the purely mathematical whiz-bang things a good quantum computer can do. How it can perform calculations 100 trillion times faster than any super computer (fact), how it can crack codes in seconds that would otherwise take years or decades, and so on. You know. The bells-and-whistles stuff.

But if we can perfect the ability to render such simulations in these new machines, the biggest contribution of future quantum computers may well be as a tool for the greater understanding of the world in which we live.

Efficiency vs Demand

That "More Power!" thing? Turns out we're not just talking computing capability. Our computing demands require actual, real, electrical power. Lots of it.

The more obvious electricity-needing things in our environment tend to draw our attention. Cars. Toasters. Elevators, drills, dryers, air conditioners and everything else that surrounds us each day; many of the things making our lives easier have quite obvious power demands.

Our phones, however, our computers and TVs, mostly just seem to be sitting there. Sure, the screens give off a little light, but that can't be much, right?

Add it all up and it is. Quite significant, actually.

Our data centers alone, the massive places that house the Googles and Apples of the world, account for up to 2% of the world's electrical demands. About equal to the consumption needs of the entire British isles.

And that's just data centers.

To not only maintain but advance the digital lifestyles we not only enjoy but demand, we need more power, not less.

Efficiency is part of it, and for many decades efficiency increased quickly and steadily. A measure called Denard Scaling, kind of like Moore's Law, measured this reduction in power consumption, and for a long time we've been able to do more with less. However that isn't a sustainable progression. Two factors are in play:

1. Increases in efficiency have slowed dramatically and are getting harder to come by.
2. Our demands have shown no signs of slowing.

Going back to our commentary in the *Power* section (Chapter III), we don't want to reduce our demands. That's wrong thinking. (Besides, who among us is *really* willing to watch less TV or use our phones less?) The idea is not to reduce what we can do but to *increase* what we can do. We want to keep demanding more, insisting on more, in order to expand this civilization of ours to the next level and the next.

Computational power is part of that.

Which means, if we're beginning to reach an efficiency threshold on the devices themselves, we need cleaner more sustainable sources for their electric power.

Yet another argument for "More power!"

The way to "go green" is not to reduce our power demands (though in the short-term that helps), but to raise our demand for the clean, sustainable power that can take us into the next age and beyond.

The Big Picture

COMMUNICATION AND INFORMATION TECHNOLOGY includes all our networks and devices, along with the slew of software and programming we run on those systems to get actual work done.

The way we interface with those systems is equally important, and big chunks of development over the years have been toward making that interface between human and machine more intuitive and natural.

Think back to the inception of computers, when programmers had to "speak" binary, ones and zeroes, in order to both program inputs and understand outputs. Our modern (non-quantum) computers, at their core, still run on those ones and zeroes—in the same way we still run on neurons and "wetware" (or whatever else powers our conscious mind)—only now, with AI, and more and more advanced programming languages built atop translators and other methodologies, we've reached a point where we can use plain language to instruct our electronic brains to deliver the results and the functions we need.

We're interfacing with our devices in the same way we would with other humans.

This is great because these are the tools that do—and will—help us both understand, and construct, our world.

Alternate Reality: Eyes On The Prize

TWO MAIN TYPES OF ALTERNATE REALITY are actively being developed.

VR and AR.

We mentioned them briefly earlier.

VR, or Virtual Reality, is immersive. This is where you put on a set of goggles which block out the world, projecting a 3D representation of a virtual world within. Throw on a pair of headphones, maybe even some tactile-feedback gloves (and/or other devices) and you pretty much escape the real world for the virtual one.

Augmented Reality (AR), or so-called spatial computing, is also done with special gear, though it keeps you in touch with the world around you. This is where info and virtual imagery is projected over what you're looking at in reality. Early examples of this were in fighter craft, using what's called a Heads-Up Display (HUD) to project info onto the canopy, directly in the pilot's line of sight. We'll begin to see this more with cars (on the windshield) and in other applications.

Both AR and VR could (and will) be used in our future space environments, from helping scientists in the lab to helping astronauts aboard space stations, spacecraft and many other applications.

The "metaverse" has become a catch-all term for these playing fields; a single, shared, immersive, virtual space (usually rendered in 3D) where we experience life in ways we can't in the physical world. Increasingly we'll do more in these virtual spaces. Those resources will help us with everything from collaboration to research to well-deserved R&R.

But ... we won't exist there.

Not for a long time, if ever.

And that's important.

(Besides, would we really want to?)

There's a saying when it comes to these behind-a-screen connections: "alone together". When we're connected online, yes, we're together, but if the power goes off ...

We're alone.

Point being, in-person is where we truly connect, where the synergy is the best, and where we need to be. It's one of those stupidly "duh" things we nevertheless forget. It's important because we're wise to remember our existence, our *actual* existence, is right here, in reality. Meaning it's vital that we advance the spaces where we *do* exist, and will exist, equally and in parallel with the virtual.

We use the virtual world to augment the real one. In concert with reality, those tools are and will be a huge boon to our progress.

As we reach further into the micro (small scale) we should, at the same time, be reaching further into the macro (large scale). Things like augmented reality, quantum computing and all those amazing things are great—they're vital contributors to our productivity, our entertainment and so on—but they're not a long-term solution to our existence. If we allow ourselves to drill in exclusively on those small-scale successes at the expense of expanding our greater reach, we lose in the end.

We've seen real-world examples of this.

As a case-in-point, in the time since the first space race kicked off we've increased our computing power by a factor of, literally, millions. In that same time our rockets pretty much go as fast now as they did then.

In fact, why are we still using rockets at all?

It's funny, back then, during the first space race, our vision of the future was warp drives and transporters, with an interplanetary society powered by the most comical computers imagined.

Remember the reel-to-reel dealios in the sci-fi shows?

They'd be flying their saucers to other planets, running computer data on tape.

And how about the flashing lights of the 1960s *Star Trek* computers? Stick in a plastic card from that multicolored stack, then listen to the mono-tone female voice after they'd ask it a question:

"Working." Repeat; flashing, colored lights: "Working." Pull out the card, stab in another. "Working."

Turns out we got it backwards. Our computers now compared to then are god-like. Beyond the imagination of the sci-fi writers of that age. Our big stuff ...

Our big stuff these days is the stuff that's comical. Compared to the way *Star Trek* imagined the future, we've failed miserably. Our big stuff, how we get around, is (mostly) unchanged since the show was filmed.

No transporters. No warp drives.

We still be launching rockets.

Driving cars on the ground.

And yet our computers are like nothing they could've dreamed.

Where did we go wrong?

So let's get back on track for the warp drives and such. Computers and communications are great, and of course key to everything, but it doesn't end there. Advancing one without the other won't help in the long run.

Balance, attacking both, is at the heart of our space future.

—

And we *are* attacking. Rockets are where we're at right now. No disrespect to that. Rockets are a vital step toward our space future. We have to crawl before we walk, walk before we run, run before we fly and so on and so on.

It's just that we started over 60 years ago with rockets and let it slip. That's the only point we're making.

Current rock stars (rocket stars?) pushing the envelope to get us back into space, the only way we (at the moment) know how (with rockets), are—and will be—our saviors. Without these incredible visionaries, out front, pushing the bleeding edge, taking those steps forward, the progress we demand will never be made.

We *are* turning a corner. Other advances in the macro are certainly near. Warp drives are further away (probability-wise; barring an exponential advance), but better and better interim options will continue to make space easier and easier.

All we're saying is, *this* time, now that Space Age 2.0 is in full swing, let's keep our eyes on the prize.

This time, let's not let it slip.

CHAPTER XIII
Education

BACK IN 1989 KOOL MOE DEE (great rapper; look him up) told us knowledge is king.

We took it to heart.

Since then your authors, us, have worked to keep our minds turned toward the worlds of tomorrow, to stay up on the latest technologies (at least as far as what those technologies are), to read up on what's happening in the fields of science and space exploration, to ask questions, be curious, stay informed.

You can too.

When it comes to staying aware of, and being ready for, our space future, there are three key pillars:

1. Seeing to yourself.
2. Encouraging others.
3. Supporting STEM.

We advocate doing a bit of each.

Seeing To Yourself

IT STARTS WITH YOU. Educating yourself is something you've been doing your whole life. Sometimes it's because you wanted to. Other times it's because others made you.

Did you love every day of school?

Odds are you didn't, but that's less a commentary on your ability to learn than it is on the system itself. When you have a purpose for studying a thing, a real, clear purpose, a passion, an interest ... that thing is pretty damn easy to learn. You look forward to it. You seek it out.

Think of something you taught yourself because you wanted to.

Were you excited to dig in?

That's the real key to education.

Which means a good question to ask yourself might be:

Am I interested in space stuff?

(Of course that's what we've been trying to do this whole time—get you interested.)

It doesn't even have to be hardcore. If you find yourself intrigued by what's happening on the frontiers of our reach into space you've got the foundation you need.

And hey, right now you're reading a space-themed book.

(See? You're already educating yourself on space.)

Hopefully it's been fun so far. You're over half done. If you got anything out of any of it up till now, you're most likely more aware of our space future.

To keep yourself educated you need only follow a few areas, a few leaders in the field, check in now and then on this or that, maybe get yourself on a few subscriptions—depending on how far you want to go.

In this day and age of information availability it's not hard to select out the things you want to stay up on.

Appendix A covers a few ways you can do exactly that.

Encouraging Others

HOW ABOUT YOUR FRIENDS? Sharing what you know is the next big step. Followed by encouraging others to do their own looking.

One-on-one conversations, group hang-outs, parties, online discussions. Find like-minds to discuss things with. Your authors call it Big Think. Synergy. This is the way ideas come about. Watch a launch together, study for a credential. Our site, FortySuns.com, is pretty new as we write this, but we hope to make it a hub where you can make such connections.

There are lots of ways to make mention of the cool stuff you want people to know about.

Our favorite: memes.

Whether in the form of something funny or enlightening, whether sent to a single friend or a group, whether posted broadly to a social channel or otherwise, memes are a great way to make people laugh and get them thinking.

The best memes do exactly that.

Would it be horrible if you became known as the guy who was always posting that dorky launch-schedule stuff?

People would soon have it on their mind, even if only at the periphery. Some might even start tuning in.

That's just an example, but you get the idea.

DID YOU KNOW?
On average we send 20B+ texts globally each and every day.

That's not posts on social channels, or even app-based texts, such as WhatsApp (*50* billion WhatsApp messages get sent each day)—that's just plain old texts being sent on our phones.

That's a lot of messages.

What if we, the enlightened, said a few things about space in there somewhere?

You definitely don't want to be preachy. No one responds well to that. But if you're sharing some cool fact or a humorous take on a current science event, it brings the fact or event to the other person's attention without you taking any position whatsoever.

It simply makes them aware of it.

And for what we're talking, simple awareness is most of the battle. When people know something about a thing, even a very little, it becomes that much more real.

Depending on your ambition, with the communication channels available to pretty much everyone today (we all have a soap box if we want it), you can go as far as you dare. Become an advocate. Start a show. Tell the world what's happening in Space Age 2.0.

Or maybe you're already active toward a cause. Maybe you already have the mad skills, and resources, to reach a wide audience and push a new message. Or add a message like this to your own.

Who you reach, how many and in what way, is entirely up to you. We ask only that you reach them.

—

Real-world example. A guy interested in rockets founded a channel called *Everyday Astronaut.* No special credentials. Just him and a passion. He's now a go-to source for everything rockets. He's living his dream, all while driving interest in our space future.

Point being, take steps.

Good things do happen.

Supporting STEM

SPECIFICALLY EDUCATION AIMED TOWARD increasing the understanding of space and the science and technologies needed to get us there.

Only ... how do you *support* education?

Good question.

If you're of means there are certainly ways to donate.

You might be an alum, with an interest in giving back to your school. Why not focus on science? A new stadium project is great, and will add to the value of the school, attract more students, add dollars to the available budget and so forth, but how about a direct donation to the physics department? Maybe there's a new laser interferometer they need in the lab ...

At the grassroots level you can encourage your kids, go to their school science fairs, find genuine interest in what they're doing, help them think outside the box.

Launch (badump bump) a local rocketry club.

Nothing would stop you from creating your own grant program. That's a pretty huge undertaking, but if you wanted to provide a boost for others that's one way to go about it.

Forty Suns would like to do that one day.

What kind of programs do the universities in your area offer?

Knowing those, getting involved in fund raisers for the programs you want to support (or coming up with your own), would also be huge.

Become a teacher.

Wear some gear for a school that delivers the education you value, with the programs you support.

Again, simple awareness goes a long way.

Supporting and encouraging higher education, the people that are exploring the bleeding edge, at the university and other levels and wherever they may be found, the curious, the bold, the determined, those advancing our understanding ...

Anything you do in that direction has the very strong potential to yield positive results.

Did You Know?

The University of California, Berkeley, along with NASA's Ames Research Center, have announced plans to build a $2 billion space center, which will become a training ground for the next generation of space professionals.

From engineers and explorers to business leaders in the fast-growing private space industry, the new Berkeley Space Center will become an epicenter for advancements in diverse fields, ranging from space robotics and planetary science to automated aviation.

There's no doubt we'll be seeing more such massive institutions coming down the line around the world, dedicated to the pursuit of space.

—

Education, for everyone, is an individual thing. We each learn at our own pace. We're each interested in the things we're interested in, and we each bring our own unique capabilities to the table.

The important thing is that we're learning.

If we may channel our inner Old Guy for a moment:

"A day without larnin' is a day wasted!"

Note, learning doesn't mean simply reinforcing currently held beliefs with new takes on the same song. There are enough echo chambers out there, and they don't tend to make us any smarter. Let's explore new terrain; move past fixed ideas, where they hold us back. Be willing to make a few mistakes. That's often the fastest way to learn.

Be willing—and determined—to find truths and apply them.

We achieve in proportion to what we attempt. If you've never made a chili before, don't make a pot. Make 30 pots over the next year. We guarantee by the 30th you'll be complimented on how well you make chili.

(Note, one of us, your authors, through just such a process, developed what may well be the greatest chili in human history. He pretty much doesn't know how to cook anything else, but dang that chili is the food of the gods.)

Whether by attention to, direct support of, or maintaining awareness in fields aligned with the pursuit of space, when we make education a centerpiece of our daily lives we do our part to ensure our space tomorrow.

CHAPTER XIV
Governance

IN A SPEECH BEFORE THIS BOOK went to press, the president of South Korea had these wise words to impart:

"In the future, countries with a space vision will lead the world's economy and will be able to solve the problems that humanity is currently facing."

This is a great example of what it means to embrace Space Age 2.0. South Korea, known for their tech companies but not really for their space programs, is making space a priority.

Many others are too.

As we move toward this new future of global participation in space—eventually becoming an extraplanetary species—how we govern, both ourselves and our fellows, will become as important as our technology and our science.

New worlds equal a potentially new canvas. A chance to apply the wisdom (ahem) gained after successfully running this one for thousands of years.

We humans *have* been wise, haven't we?

Surely by now we've figured out what works best?

Distances will impose constraints. New worlds, even outposts and bases, will need ways to self-govern in the absence of easy remote oversight. Help, military action and law will become matters of local rules and regulations.

What seeds are being sown?

Treaties & Accords

SOME FIFTY-PLUS YEARS AGO, in 1967, we drafted—collectively, as a world—an Outer Space Treaty. Future historians will note this was also the year one of your humble authors was born.

Can you believe people were already thinking space governance all the way back then?

It's okay. I (Dave) know I'm old.

But it illustrates that humanity has been planning for a future in space for quite a while. Since the time of the first space age. We take this as a good sign. This generation and the next will need to go further, likely establishing local governing pacts for remote bases, installations, and other farflung places in our solar system, but at least we have a foundation.

The Outer Space Treaty is a multilateral pact that forms the basis of international space law, outlining, among other things, agreements:

- Prohibiting nuclear weapons in space.
- Limiting the use of the Moon and all other celestial bodies to peaceful purposes.
- Establishing that space shall be freely explored and used by all nations.
- Precluding any country from claiming sovereignty over outer space or any celestial body.

As of 2023, 113 countries were party to the treaty, including all spacefaring nations.

It's in need of updating, as noted. Current generations will need to take stock of the incredible advancements made since the Treaty was first drafted, as well as accounting for the diverse players now involved (private companies, as well as nation states), in order to come up with new, better guides that work.

We're confident it can, and will, be done.

The Artemis Accords

IN 2020, drawing from the Treaty, the Artemis Accords were born. A non-binding multilateral arrangement between the United States government and other world governments participating in the Artemis program. The Artemis program is an American-led effort to return humans to the Moon by 2025. The ultimate goal, of course, being the expansion of space exploration to Mars and beyond.

The stated purpose of the Accords is to "provide for operational implementation of important obligations contained in the Outer Space Treaty and other instruments."

Fresh signatories have been rolling in as we work to get this edition ready for publication. We can't keep up! Peru and Slovakia are expected to sign, which will bring the number past 40.

Ten signatories ago it was Bulgaria; before that the Netherlands, Iceland, Germany and Argentina, before that the Czech Republic and Spain, followed by Ecuador and India. (FYI, India is one of the world's most advanced space powers, just behind the United States and China.)

Since Bulgaria it's been:

- Angola
- Belgium
- Greece
- Uruguay
- Switzerland
- Slovenia
- Lithuania

Now Peru and Slovakia.

By the time you're reading this we hope others will have pledged their commitment to these unifying Accords.

DID YOU KNOW?

The US Commercial Space Launch Competitiveness Act, passed in 2015—also referred to as the Spurring Private Aerospace Competitiveness and Entrepreneurship (SPACE) Act of 2015—allows US industries to engage in the commercial exploration and exploitation of space resources.

This is an example of lawmaking and regulation acknowledging the needs of the growing space industry. Increasing efforts to consolidate regulations is a good thing. It means the industry is being taken seriously.

The Act extends "learning period" restrictions through 2023. It has already been extended. This learning period was put in place to limit the ability of the Federal Aviation Administration (FAA) to enact regulations regarding the safety of spaceflight participants, while it gathered data.

That period has now come and gone, and the FAA is working on updated rules for non-government astronauts (those not trained by NASA and employed by the US or a partner government), clarifying how things will work going forward.

Another great sign of just how real space has become.

Keeping Pace

SPEAKING OF THE FAA, and launches here in the States in particular, pacing becomes a factor that must be addressed. Gone are the methodical delay-filled expectations of the last space age. That pacing will no longer work. The gap between the speed of our governmental launch regulations, such as those overseen by the FAA, and our launch cadence itself, threatens to slow progress if not adjusted and enhanced. As commercial launch demands increase, rapidly, with launch rates literally skyrocketing, wait-times for approval processes and other factors that put the brakes on "getting it done" must be eliminated or, at a minimum, streamlined in order to keep up.

Bill Gerstenmaier, vice president of build and flight reliability at SpaceX, says: "The pace of American regulation must match the pace of American innovation."

This isn't to say regulation must be ditched. Quite the contrary. Oversight is imperative. Safety.

The system just needs to be beefed up.

Significantly.

Industry should never have to wait on government.

Referring back to an earlier portion of this book, where we discussed how much mass will need to be lifted into orbit as we accelerate our presence in space (the part where we talked about launching tanks into orbit? Yeah, we thought that might be a better reminder), in order to meet those requirements we'll need not just the technology to get it done, but the approval and safety processes as well. Those regulations must be in place and well thought out *before* we need them.

We can't afford to trip-up the rapid expanse of our commercial industry. Red tape, bureaucratic slowdowns ... those things have the potential to slow meaningful progress.

These systems must work in harmony, they must advance in parallel, as one. Regulators must share the same mindset, the same urgency of those pushing our reach into space.

Safety. Speed.

Zero delays.

We have no time for time.

No one, at any level, should ever be waiting.

Check it. Fix it. Check it. Launch it.

Things have to *move*, and move fast.

Meaning our regulatory bodies need to "beef up to keep up".

One of our strengths as humans is the ability to evaluate the future and make plans. It's how we have civilizations, train schedules, tiered prizes for ugly sweater contests, elections ...

You get the idea.

We look ahead. We estimate what will be needed. We prepare.

Not always perfectly, of course, sometimes quite horribly, but the fact that an Empire State Building exists in New York City—or that a New York City exists—would seem to speak to the notion that, mostly, we're pretty damn good at figuring out how to do stuff.

Which brings us to our point:

The Empire State Building was built in 410 days, start to finish.

Think about that.

We *can* pull off massive projects with super speed.

There's no reason things need to move slow in order to achieve history-making results.

Bureaus, Councils & Departments

GOVERNMENT AT EVERY LEVEL will help organize and provide oversight for our space future.

This entire area is evolving. Rapidly.

Over the last many decades we've formed, dissolved, and formed again numerous agencies for the management and pursuit of space. One in particular, the National Space Council (NSC), has been killed and revived twice. Established in 1953 as the National Aeronautics & Space Council (NASC), it was ended in 1973, only to be reborn in 1989 as the NSC. Subsequently disbanded in 1993, it saw new life in 2017, when it was revived with the purpose of having "a significant involvement in the direction of America's activities in space." So far it's proven a good guide for US space policy.

And check this out. In late 2023 FCC Chairwoman Jessica Rosenworcel revealed a plan to split the International Bureau of the FCC into two distinct offices:

- The Office of International Affairs
- The Space Bureau

Say that once out loud:
The Space Bureau.
Has a nice ring.
Ms Rosenworcel says, "A new Space Bureau at the FCC will ensure that the agency's resources are appropriately aligned to fulfill its statutory obligations, improve its coordination across the federal government, and support the 21st century satellite industry."

Another sure sign we're well into the next space age.

These sorts of hierarchies and new organizational forms will become more and more commonplace as we build out our governing infrastructure.

State Players & More

USED TO BE THE SPACE RACE was almost exclusively between the US and Russia. Lately a new order is emerging. China and the United States are now the biggest competitors. Such space races mean more rapid advancements across the board. Yet, this hearty (and yes, often contentious) competition still won't be enough. The demands of space are high. Fortunately other state players continue to enter the race, along with a huge number of private enterprises.

This is good. Competition, whether national or corporate, moves the needle. As well, your support, your fandom, helps. In fact it's vital. Because even with those two great superpowers in furious competition, plus a bunch of other State players, along with more and more companies trying to make money, they *still* won't be able to do it without the rest of us.

Not to the degree that's really needed.

True success in the arena of space will take shoulder-to-shoulder, global intent. A space race against ourselves, if you will, driven by the demands we ourselves impose.

This includes each of us.

Such wide participation at all levels has the benefit of keeping everyone on-task. If one player decides it's too expensive or too much trouble, another will step up and get the prize.

Pressure.

Good pressure.

As a recent example of this, under a bilateral agreement with NASA, the Italian Space Agency (ASI) and Thales Alenia Space (a French-Italian space manufacturer) are collaborating to create the first permanent human habitat on the Moon.

Way.

This year (as we're editing this section), Thales announced a contract with ASI to develop what is being called the Multi-Purpose Habitat (MPH)—a key component of NASA's Artemis program. As you recall Artemis aims to establish a sustainable long-term human presence on the Moon.

This joint project to develop the MPH could conceivably result in the first permanent Moon base.

Yet another way we're witnessing this "competitive cooperation" paradigm evolve. With such collective intention, with focus from each of us, at all levels, this time we won't fail.

Did You Know?

China has space contractors, just like the US.

China Aerospace Science and Technology Corporation (CASC), the country's main space contractor, recently outlined a series of goals, including a crewed lunar landing along with other exploration and transportation goals—while stressing the importance of space infrastructure and developing capabilities such as on-orbit servicing, and building a space governance system.

The overarching ambition is to make China one of the world's main aerospace powers by 2030, becoming a fully comprehensive space power by 2045.

PUSHING THE ENVELOPE

NASA has made it clear the agency's goal is a low Earth orbit marketplace where NASA is one of many customers, and the private sector leads the way.

The purpose of this strategy is that doing so "will provide services the government needs at a lower cost, enabling the agency to focus on its Artemis missions to the Moon and eventually Mars, while continuing to use low Earth orbit as a training and proving ground for those deep space missions."

Eternal Vigilance

AS OF 2019 the United States of America founded the world's first and (so far) only dedicated independent space force.

On December 20th of that year the US Space Force became the sixth branch of US military service (Navy, Army, Marines, Air Force, Coast Guard, and now, Space Force).

As of this writing, the Space Force has 8,600 service members and 77 spacecraft.

Sounds like a fully-formed branch to us.

Also, the Pentagon recently announced the Space Force plans to buy even more rocket launches from companies in the coming years than previously expected, granting more companies a chance at securing billions in potential contracts.

If this doesn't put an exclamation point on the fact that our future is in space, not much does.

One way or the other, we humans are committed to pushing into that Final Frontier.

An Interesting Family History:
Way back when, there was only the Navy and the Army. From the Army evolved the Air Force, and now from the Air Force has evolved the Space Force. Making the Army kind of like the grandparents, with the Air Force appearing in the holiday photos as Mom and Dad.

DID YOU KNOW?
The Space Force is 'pushing hard' to allow more sharing of information with allies.

Deputy chief for operations cyber and nuclear said a growing number of spacefaring nations are looking to work with the United States on technologies and strategies to defend their assets from anti-satellite weapons, and are increasingly making their own investments in space defense systems.

On that note the Space Force is ramping up efforts to work more closely with allies. This is a top priority of the chief of the Space Force, Gen. Chance Saltzman, who declared "partner to win" as one of the service's ideals.

PUSHING THE ENVELOPE
Code name VICTUS NOX. That's the US Space Force mission that was designed to test the readiness response time for a rapid satellite launch.

Two companies were selected, Firefly Aerospace and Millennium Space, after which they were on "hot standby," awaiting an alert notification from the Space Force. After getting the alert, the companies had a 60-hour window in which to transport a payload to the launch site, conduct fueling operations and integrate it with the rocket. They pulled it off in 58, not only beating the requirement, but handily destroying the typical timeline for such a maneuver of weeks or months.

After that the Space Force gave the final call to launch, along with orbit parameters. Firefly then had 24 hours to stand ready to launch at the first available window.

"Liftoff took place at the first available launch window, 27 hours after receipt of launch orders, setting a new record for responsive space launch," said Lt. Gen. Michael Guetlein, commander of Space Systems Command.

So, basically, orders were given to launch a satellite into space and, 3 days later, it was in orbit. From total "cold iron" to an orbiting, functional satellite.

Did we say space operations are getting real?

Apologies, but we just can't say it enough.

The future is now, baby.

—

In addition, that same Space Rapid Capabilities Office (Space RCO), a Space Force agency meant to provide technological solutions to military problems on a fast timeline, is, for the first time (as of this writing), sponsoring the annual Hyperspace Challenge. We mentioned this briefly back in the "Moving in 3D" section in Chapter IV.

Originally meant to find small businesses looking to break into the defense market, this year's Hyperspace Challenge will be looking to connect existing, later-stage companies and their technologies with USSF needs. The idea being to "streamline the design, development, and deployment of a solution within the next few years."

(Shh. Don't tell anyone we said so, but this space stuff is getting real.)

Oh, and for pure (yet so cool) trivia, the USSF emblem has very definite, and deliberate, echoes of *Star Trek's* Starfleet logo.

Take a look when you get the chance.

The Best Of The Best

BACK TO GOVERNANCE. After millennia of conquering each other, coming up with exciting (and not so exciting) ways to rule, we humans have piloted a few successful methods of governing on this planet we call home. The ideal of large groups of people living together in happiness (as much as possible) and harmony (again, more than less), working effectively, productively, accomplishing shared goals together, is probably one of our best.

Not every way of governing has shared that ideal.

But the best of them have. Whether any have managed to pull it off fully or not, the effort has been there.

Could there be other/better ways?

Those experiments and successes (built on learning from failures, of course), while the product of millennia of evolution, don't necessarily represent all the options there could possibly be. We have to believe fusions of the best of the best might lead to an even brighter future.

Who knows? The best form of government ever imagined might be dreamed up tomorrow.

In the same way technology advances, so, too, does the way we treat each other, including how we participate in common pursuits. Leading and being led, both of which require unique forms of willingness, are tough to perfect. Open dialogue is key. The scientific world enforces this with rigor. All are allowed (and expected) to speak, to present their case, their findings. All findings are then considered, evaluated, and ...

Something interesting happens.

Progress is made.

In order to truly move the needle on progress toward our space future, we can't afford to forget this lesson from science.

Debates have already begun, and are escalating. Which is good.

How do we prep for the near- and long-term future? The groundwork we lay now is vital. Rule of law, mining rights—the topics that will need to be addressed are many.

Decisions will have long-lasting consequences.

Philip Metzger, a planetary physicist with the University of Central Florida, said, "I do not believe this is a time to coast along carelessly. This is a tipping point in the trajectory of human civilization. We must be intentional to set precedents with the focus on their 100-plus year consequences."

Hear hear.

Make Your Voice Heard

OUR GREATEST ADVANCES IN CIVILIZATION have come during periods of our most open dialogues. We stunt our growth in direct proportion to the voices that are not allowed to be heard.

No government, no entity, has life, no ruling form works, without strong, free-thinking individuals. People make up any larger unit, and good people make that unit strong. Any government, or any group, is nothing, quite literally, without its individuals.

Get involved, when and where you can. Make your voice heard. Together we steer the course of our future, from votes on funding to board decisions to all else that shapes our actions, both small and large.

Our space future depends on each of us doing our part.

It depends on us being willing to talk.

To listen.

To learn, together, from our failures.

To celebrate, together, our wins.

(And if we're smart, we'll learn from those, too.)

—

Seventy-two countries now have their own space programs, mainly because they can't afford to be left behind. Yes, that probably means the motivation in most cases is FOMO (Fear Of Missing Out), but we'll take it.

The important thing is they're participating, and are driven to make space a reality.

Which is a good thing.

Our progress depends on it.

CHAPTER XV
Progress Is Possible

"You will either step forward into growth, or you will step back into safety."

— Abraham Maslow

AS YOU CAN SEE, lots is already happening in space, with lots more either planned or underway. We just spooled out what felt like a rush of facts and info, tying it together the best we could to paint a picture of Space Age 2.0. Our sense of it so far—even though we're the ones who wrote it—is almost like sitting trackside at a Formula 1 race. A lot of really flashy, really cool, really impressive things just went whizzing by in a blur and, while we saw a lot, we mostly don't recall any details. Like the race, however, when it comes to space we're pretty sure of two things:

1. There's way more happening than we thought.
2. It's pretty damn exciting.

And so here we begin a look at some of the ways we'll ensure this new space age is here to stay, along with what achieving that reality will involve.

Let's take another look at our opening mission.

To unite the focus of humanity toward common extraplanetary goals that create an ever-improving future, by building upon the things we hold in common despite the things we hold different.

Doing that will take cooperation. Cooperation, common goals, this is what *Forty Suns* is all about. These final chapters, then, including several, worthy appendices, are meant to give you a few ideas as to how we can attain our space future.

A Basic Premise:
Our world is an awesome one; it can also be a troubled one.
By being willing to look directly into the face of the obstacles, and the people, that would thwart human progress, we soon see how weak those contrary positions truly are.
Not only is progress possible, it is inevitable.
The Borg said it best:
Resistance is futile.

A Sprint, Not A Marathon

THAT'S USUALY WRITTEN the other way round, "a marathon, not a sprint", and while we don't disagree that the road ahead is long, with lots and *lots* to do, many miles to cover and so on, we think the "race to space" humanity is poised to undertake is better framed as a series of sprints.

Call it a relay race. A long one.

In much the way a baton is handed off from runner to runner, each runner giving it their all, viewing progress toward our space future in discrete units is a more accurate analogy.

The race will be long. Each racer will be moving flat-out.

Maximum effort, to quote Marvel anti-hero, Deadpool.

Speed is of the essence, but it will also be a collaboration of speed, making each runner vital.

Each program, each scientific discovery, each new mission—in each case we move fast and learn as we go, fall down and get up, innovate and iterate, each of us sprinting when it's our turn, such that, collectively, the relay race is won.

In this way rapid progress is made.

DID YOU KNOW?
In the interest of organizing a segment of this larger race, NASA has established the new Moon to Mars Program Office (mentioned earlier), to carry out the agency's human exploration activities at the Moon and, eventually, Mars.

According to NASA, "The Moon to Mars Program Office will help prepare NASA to carry out our bold missions to the Moon and land the first humans on Mars. The golden age of exploration is happening right now, and this new office will help ensure that NASA successfully establishes a long-term lunar presence needed to prepare for humanity's next giant leap to the Red Planet."

We like the sound of that.

Especially the part about: "The golden age of exploration is happening right now."

—

What can we be doing? So many things. The biggest is to stay interested, stay informed, be supportive, talk about the latest this, the coolest that with our friends ... basically, be a fan.

Does participation require you to be a scientist?

No.

Does it require you to work for a space-technology company?

Definitely not.

And that's what makes *Forty Suns* different. We're actively encouraging you to simply fly the banner of space.

Of course, the net result of that is that more of us *will* get involved. In the same way progress is inevitable, if more and more of us are getting fired up about space, at least a few will parlay that keen interest into greater participation.

Did You Know?
Once a year, NASA scours the field for the most out-there, far-fetched inventions it can find with the potential to transform the future of spaceflight, and awards them funding.

It's called the NASA Innovative Advanced Concepts (NIAC) program, and its purpose is to nurture visionary ideas that could transform future NASA missions.

In the latest round six concepts were selected for continued study. Among them a sprawling radio telescope array on the far side of the Moon that could reveal new secrets about the universe, along with a concept for a way for astronauts on long-duration missions to grow their own medicines.

Any budding space innovators out there?

The Tipping Point

Where do we cross the line?

Where do we, as a civilization, cross without question into the realm of the extraplanetary?

When do we reach and go beyond the point where we become a global, cohesive force toward the conquest of space?

And what does that mean in terms of lost autonomy?

As with all things, it's important to look at both sides of the coin.

We've been pitching for more global cooperation when it comes to space, but what's the downside? Do we sacrifice too much if all of us are engaged in the same pursuit? Can we truly work shoulder to shoulder, as a planet?

Where do those of us who *don't* agree turn?

What if we don't *want* to be part of a space future?

If the whole world is doing it ... how does that work?

Globalism is often a swearword. We're not advocating extreme globalism, but bringing all of us together toward a common cause leads in that direction. Philosophical debates, about this crazy, mixed-up world unifying toward a common cause, might rage.

Is the risk of homogeny (and yes, global homogeny without options could be a risk), even if only temporary, worth the reward of brave new worlds?

Food for thought.

War! Good God Y'all!

ANOTHER WAY FORWARD would be to battle our way into space. Fight against each other even as we innovate independently, in a sort of winner-take-all scenario.

That might be the way it goes.

Truthfully, there's a good chance that *will* be the way it goes.

It might also not be the way we want.

On the positive side—if there is a positive side to such a scenario—there's the possibility that, in an intensely combative, competitive environment, we'd accomplish some of those bigger goals faster.

Maybe.

Or we could just blow the whole thing up.

Swing the pendulum the other way.

Give up.

Whatever.

May be a while before we find another world like this, though. And when we do, it would be a long haul to get it back to where we are now. Of course that's assuming you believe in a life continuum. One where we could annihilate ourselves here, then pop up on some far-distant world in some far-distant future.

Seems not only a risk but a waste.

And if that's *not* the case, if we truly are glorified mud, if it's dust to dust after all, if death is the absolute end and all that, well ... we wipe the playing field here and that's it.

Game over, man.

So maybe annihilation isn't the best option?

In either of those realities.

Being open, however—which is what we're trying to do—does require considering all possibilities, grim or otherwise. Fighting is in our nature.

Fact: It's been a long time since the last truly global conflict and the world is getting antsy. If there was ever a time to turn our attention outward it's now. Doubly so because now, unlike at any other point in history, we actually have something external to *all* of us on which we can legitimately focus. A very real, very attainable goal (space, in case you forgot) that can become part of our shared objectives worldwide.

Absent such shared focus we might slip further, to a brink we'll regret.

Wars, some thinking goes, are the real foundation of our political systems. Of our world, our history. Which means it might not be as easy to "make love not war" as we'd hope. A larger war might be inevitable.

We humans love our army stuff.

We really do.

Don't try to mess with someone's right to wage war, it seems. Nuh-uh. Most other things are fair game; that one's off limits. Shadow players, nefarious bad guys, ne'er-do-wells and creepy, dark figures at the highest levels that supposedly run our world, if real, do so through the fabrication and manipulation of wars (and us) and the weapons that power them.

That certainly seems possible.

Maybe even likely?

If it's true, how do we change it?

Those last few sentences are probably wrong on many levels.

Point is, even if true, even if war *does* underpin our entire existence, that's about the dumbest agreement we could make as a species, don't you think? Especially at this stage. Wars here on Earth are tiny games on an average planet in an average galaxy in a tiny corner of what may, truthfully, be a tiny (and possibly only one of many) universe. Nothing games on a Nothing world.

Yet those games have the ability to end us.

Or at least slow us down.

Yes, these 'games' are in our face. Not just wars, but all the petty obstacles to progress we invent day in and day out. Large and ponderous systems that seem quite complex, filled with overwhelming inertia, quite serious, seemingly intractable ways of doing things that drive us to anger, despair.

Yet they're games of our own creation.

Any system thwarting us on this planet is here because we put it here. Don't send for whom the bell tolls; it tolls for us. We're the architects of any and all conditions befalling us here on Earth, good *or* bad.

Meaning we're subject to obstacles of our own creation.

The idea of viewing the sociopolitical machinery of our world as petty games may seem crazy, but isn't it more crazy to keep playing them? Didn't we decide to play them in the first place? Could not we, as a group, decide to play different games? Maybe at least *one* different game? Perhaps the pursuit of space?

As Plato said, only the dead have seen the end of war, and that may be true. Yet ... a little break might do us some good.

Hm.

At its core that oversimplified idea may not be as naïve as it sounds.

This is our playground. Us humans, here, on Earth. It belongs to us. Children get fussy, yes—it happens, even to the best among us—but we're it. There are no adults waiting to come break us up and make us play better games. We're the children, and the adults, and if we don't fix ourselves no one will. The playground has gotten pretty crowded; if we want more places to play we need to work together. That includes the "stoppers" (people who obsessively stop things from happening), the bullies, the bad guys, the doers (god bless them), the saints and the sinners. The good, the bad *and* the ugly. All of us.

No one else is going to call a timeout.

No one else makes the rules. We do. Anything happens on this planet, when it comes to human affairs, happens because we say so. Because we allowed it to. Either because we wanted it, or because someone with mal intent wanted it and we failed to stop them.

Good thing is, it's never too late to change our minds. If we want better, let's make better. Competition, not conflict. Maybe we start there. A little shift.

Healthy competition rather than hostile conflict.

Both can strengthen a system. The fallout from conflict, however, means damage that must be undone. Setbacks. In competition, on the other hand, even the losers are part of the overall win. More sports examples: Rugby typically has a "third half", where opposing teams get schnockered together after battling for blood in a game of sport. Most MMA fighters embrace after the fight. Football players shake hands. There's no reason that spirit can't scale. No reason global institutions can't benefit from the exertions of high-level competition, deriving camaraderie from that very effort.

The One Tenet

THE CONTEMPLATION OF SUCH A THING is not missing the bigger picture. In many ways it *is* the bigger picture. Honestly, that's what the world needs more than anything right now: Less complexity, not more.

It *is* simple.

Be nice.

Have a care for your fellow humans.

Apply the Golden Rule. Did you know that's the *one* tenet common to *all* religions, beliefs, betterment doctrines and anything else with an eye toward guiding man to a higher existence? That one message? Around the world, throughout human history. A principle so common, so sane, it *is* us. No single wise man dreamed that one up. It's woven into the very fiber of our being, part of our existence since before the beginning, coming forward with us throughout time.

We ignore it at our peril.

Worded differently in some cases, the message is the same:

Do unto others as you would have them do unto you.

Simple.

It takes only a decision.

Ideologies and other made-up stuff be damned.

We're human first.

An ideology or a political belief should never stand in the way of getting along. Of cooperating.

These days we're too quick to anger, too quick to find fault, and it's not really our native frame of mind. It's mainly a product of the artificially contentious atmosphere in which we live. We weren't like this to each other a hundred years ago. Even fifty. We gave each other the benefit of the doubt first. Being nice was the default. Kind of like "innocent until proven guilty", we were "understanding until you give me reason not to be".

Our default these days seems to be to expect an argument first.

These are other people just like you. Whether online or live and in person, these are other "yous", with the same foibles, the same flaws, the same potential to err, the same desire to be right, the same reluctance to admit to being wrong.

The same goal to live and let live.

Treat them accordingly, until you have reason to do otherwise. Expect them to be human first, like you. Let them prove to you they're an asshole. Until then be nice.

You'd like that if people did that to you, right?

Instead of yelling out the car window at someone who just cut you off, maybe give them the benefit of the doubt first.

You're not an asshole (we'll give *you* the benefit of the doubt on that one, and we're very likely correct), but have you ever made a mistake? Accidentally weren't paying attention and maybe cut someone off? You're not an asshole, worthy of a shout. You *are* worthy of some understanding.

Life can be a vicious cycle or it can be a positive one. We, each of us, choose which. Decisions are powerful things. We can choose to put the brakes on the negative. Stop it in its tracks and turn the flywheel upward, toward a better future.

The Bigger Picture

WHEN IT COMES TO SPACE, how about this scenario instead:

Rather than acting like it's a big deal, maybe we just politely ask some of those important people, where they exist—you know, the ones we've agreed to let be important (they're only important because we say they are)—to give it a rest long enough to let us get the show on the road? To look out, not in. While at the same time the rest of us agree to be more civil to each other. More decent.

At least long enough to become extraplanetary.

After that we can go back to blowing things up, if we must. Being mean, if we choose. Waging wars. Imagine our own future version of Atreides versus Harkonnen (warring Houses in Frank Herbert's sci-fi epic, *Dune*), with vendettas spawned over disputes involving entire star systems.

Now *that's* a game.

For that to happen we have to *get* there.

Our current war-hawks may be big shots (again, because we let them be), but the truth is they're just kids on the same playground as the rest of us. For them we have a simple message:

Knock it off.

Be better.

"But it's more complicated than that! We can't let the other guys get the advantage on us! We *have* to fight them and destroy them before they destroy us and our way of—"

Sure, sure.

Maybe a little whack on the back of the head, a quick pop so the ones calling these shots wake up and see what idiots they're being.

("What's wrong with you!")

There's too much to gain, and too much to lose.

Let's hold them accountable or get rid of them.

The decision to make change is one of our super powers.

United We Stand

WE MENTION THESE THINGS only to draw attention to the fact that increasing global collaboration will bring with it resistance and debates, none of it likely grounded in anything other than the reactions of special interests that are being threatened. In the same way there are negative people among us, there are larger, vested interests scurrying through the halls of power that might not share our enthusiasm.

Remember, there are advantages for some to keep us divided. To sow division. Others win when we're fighting among ourselves. For them it's better to pick at scabs and not let old wounds heal. For them, it's better that we fight. For them, it's better that we bleed.

Such conflict raises no one.

Those people are not our friends.

We're a world of problem solvers.

Fact:

There are a lot more of us than there are of them.

What about when it comes to the subject at hand?

"Why would anyone want to stop us from going to space?" you ask. "That would be stupid," you say.

Indeed.

It's one thing to simply not pay attention or just not care. That's bad enough. But to actively oppose the betterment of humanity as a species? That, friends, is the true face of the enemy, no matter how "reasoned" their arguments might seem.

Again, though they may hold power (in certain cases), we outnumber them a hundred to one. That figure isn't entirely random; it's pretty accurate. More often than not the ones you encounter who seem to be of that ilk have, in fact, simply bought into the assertions of one of the few, truly negative people. They themselves are not truly the enemy.

Cleared of that negative influence their own hope is restored.

Meaning, if we decide to make it so, there's nothing the small minority that would truly thwart us can do. Meaning it's up to us to ignore them and move forward.

Those who would stop us can easily be steamrolled. If we're united.

We only fail when we stop to listen to reasons why we "shouldn't" and why we "can't".

See the *Can-Do* section further down in this chapter.

Getting There

LET'S GET BACK TO THE MUNDANE FOR A MOMENT.

The mechanics of getting there.

When it comes to making progress, our capabilities define us. At the moment we have but a few active crew capsules for carrying people into space.

Four, in fact:

1. Soyuz (Russia)
2. Shenzhou (China)
3. Dragon 2 (US)
4. New Shepard Crew Capsule (US)

These are actual crew modules that can sit atop a rocket and hold astronauts as they're launched into orbit.

In essence, the above options are all we have currently for getting people into space.

Several others are in development:

- Orion (US – Artemis program)
- Starliner (US)
- Orel (Russia)
- Gaganyaan (India)
- Chinese Next-Gen (China)
- Kavoshgar (Iran)

So we're improving. This is a good sign.

And yes, you read that last entry right. Iran has not only a space program, centered by the Iranian Space Agency, but they're putting together plans for a crewed capsule.

The Kavoshgar.

And you thought we were kidding about everyone getting into space.

The whole world is doing it.

New Milestones

REMEMBER THE SPACE SHUTTLE PROGRAM? That was America's last big national effort at making getting cargo and crew into orbit mainstream.

How we loved that design as kids.

As for duration, seemed like the Space Shuttle program went on forever. It pretty much did, spanning four decades. (Five? Technically it crept into the 2010s.)

Nearly *40 years*.

Yeah. We had to check the numbers too.

- SS program dates: 1972 – 2011
- First flight: Aug 12, 1977
- First crewed flight: Apr 12, 1981
- Last flight: Jul 21, 2011

There were five complete Shuttle orbiter vehicles in the fleet, which were built and flown on a total of 135 missions from 1981 to 2011.

A lot got done during that era.

Wanna hear something interesting?

Despite that long track record, in the most recent decade, with Space Age 2.0 picking up momentum, two Shuttle records have already been broken.

Kind of a big deal.

Forty years of action for the Shuttle, and in the last ten SpaceX has exceeded two of that program's milestones:

1. Trips to the ISS, and
2. Time in orbit.

That last may be the hardest to believe. But sure enough, not only has the SpaceX Dragon capsule now been to the International Space Station 38 times vs the Shuttle's 37, on Tuesday, June 6th, 2023, the Dragon 2 fleet's cumulative time in space surpassed the space shuttle fleet's time with 1,324 total days in orbit.

Did you have any idea one company had spent that much time in space?

That's what we call getting there.

One last note on the legendary Shuttle.

When it comes to getting things done in space, function is the goal, of course, and when it comes to designing for that function, well, form follows in line. The Shuttle's form, though, was such a break from the traditional rockets of the day ...

It fired imaginations around the world. The Shuttle became an icon not just for the space age, but for the inspiration of millions.

The Shuttle had a purpose, and a mission, which it performed admirably, yet its designers may never have foreseen its other mission: the seismic shift of attention it would bring to the world, turning all eyes toward space. They may never have realized how that raised awareness would end up being one of its greatest successes.

Inspiration, after all, is the foundation of our future.

Inspiration drives interest, interest drives involvement, involvement drives progress.

As an icon, the shuttle delivered on all fronts.

We believe SpaceX's Starship will become one of the next major icons of the new space age.

—

Ooh! If you liked the Shuttle, check out the Dream Chaser spaceplane by Sierra Space, the same guys doing the inflatable space station (these guys truly are cutting-edge). This is a new, multi-mission concept vehicle designed to transport crew and cargo to low-Earth orbit (LEO), which, in their words, can be customized for both domestic and international customers for global operations.

It launches on a rocket and flies home like a plane, the way the Shuttle did—only way more flexible (and cheaper).

Two models are planned, the DC-100 and the DC-200, uncrewed and crewed respectively.

Did You Know?

Speaking of space planes, the Boeing X-37, also known as the Orbital Test Vehicle, is an American reusable robotic spacecraft and another glider. It's boosted into space by a launch vehicle, then re-enters Earth's atmosphere and lands as a spaceplane.

Echoes of the Space Shuttle can also be seen in this one. Though we've gone back to more traditional rockets as of late, the idea of "flying" into space then flying back in the atmosphere to a landing is an intriguing one, and one that will likely be a big part of our space future.

As of this writing, China has landed a similar reusable spacecraft of their own that spent nine months in Earth's orbit, similar to the American X-37. It's likely a craft with potential for research as well as military uses.

That makes China just one of a handful of entities that have successfully operated a reusable spacecraft.

Both planes are examples of a national interest pushing the envelope when it comes to space.

Can-Do

A QUICK INTERLUDE TO RE-FRAME what our space future will take. It won't be easy. We're clear on that. At times it may even seem impossible. Certainly, as we've noted, some small minority will insist that it is. To get there, in any capacity, will require the very can-do spirit that's gotten humanity this far. Everything we have today, from mass-produced stick-pins to one-off space stations, we have because at least a handful of us were driven by, and acted on, that can-do spirit.

Not can't-do.

Can't-do people are convinced nothing outrageous is possible.

"People can't fly," they might insist. "Birds fly, bugs fly, people can't fly. That's a crazy idea and you're a fool for trying."

They're all about why we can't.

Why we shouldn't.

How dangerous it is, how doomed to failure.

What's the point in being that person?

There are only two possible outcomes if you're them:

1. No one ever flies, and you get to smugly say, "I told you so." Which means you don't get to fly either. What a great life.
2. We end up flying, and you look stupid.

Both outcomes are losing propositions.

Instead of poo-pooing things, help figure out how we can actually make them real.

What if the approach of such people was instead:

"Well, I think it's crazy to try, but dang, let me think about this. How *could* we fly? Impossible, I tell you. Only, if we could somehow figure this fool thing out it would sure make life easier. Dang if I see how, though. Well, if you've got ideas let's hear 'em. I'll do my best to support you."

That spirit is so important. If you find yourself listening to someone who only wants to talk about how we can't, we invite you to stop listening. You may be losing valuable time. Progress is not about worrying over how we can't, progress is about deciding how we can.

The bigger the challenge the bigger the solution required, of course, but we don't stop looking for solutions just because things get hard. We don't stop seeking answers because someone said we can't. No matter how many "reasons" they might point to.

Not that their reasons are invalid. The things they worry over may be quite valid, in fact. No, we ignore those people because the way forward is by finding ways to solve those reasons, to do what we decide to do in spite of the obstacles.

There are and have been articles, even entire books, on why we "can't" live in space, why we "can't" live on Mars, why we can't this, why we can't that—you name it and someone somewhere at some point has gone to great lengths to explain to us why it "can't" be done.

Hogwash.

Anything dedicated to telling us why we can't do great things is advice we, frankly, don't need. Why take any of that to heart? If we listen to those people at all it would be for one purpose alone: To get the dissenting view, so maybe we spot a few things we didn't think of that we need to solve.

Can't live on Mars because of radiation?

Great. Thanks for pointing that out. Oh, you're just saying we can't but not giving any solution?

We'll come up with our own.

See? Even the negative nancies might find some usefulness in forging our brighter future.

Einstein was a bit less accommodating. His view:

"Stay away from negative people. They have a problem for every solution."

True enough. Problems are where the "can't" people live. Everything is a problem.

Funny thing is, they're not wrong.

Wait ... what?

It's true.

Everything *is* a problem, to some greater or lesser degree.

Problems *do* exist.

Lots of them. For anything.

Where they're *horribly* wrong, these doomy people, is in the fallacy of their computation. To them "problem" equals "can't".

Which is why they never accomplish great things.

That's why they never get to be in charge. Their only (semi) useful function in life is, again, to shine a light on things we need to solve.

Maybe that's what makes them bearable.

Otherwise they're not useful.

Because why?

Because the goal of existence isn't an absence of problems. Life is filled with problems. One might go so far as to say life *is* problems, the posing and solving of. In case you hadn't noticed, once you get rid of one problem new ones find their way in. The key—perhaps counter-intuitively—is to embrace them. Relish a good problem, solve it and win.

Then dream up the next one.

Which means we want to make our shift of approach less about avoiding problems, more about the nature and the scale of the problems we choose to take on. Expansion doesn't tend to happen by solving small problems. Mere existence, basic survival might work by solving small problems; not so much with rapid growth. We expand through taking on bigger problems, solving those then taking on still bigger problems, then yet bigger problems and on and on.

We grow through challenge. By extending our reach, grasping for higher branches on the tree, elevating our demand. The greater the challenge, therefore, the greater the reach, the higher the demand, the greater the growth, the greater the challenge ...

And so it goes.

Progress is inevitable. People who waste time harping on why we "can't" usually end up very wrong in the end. History is absolutely overflowing with such examples; people who said we couldn't and then we did. "Getting there" would never happen if anyone ever listened to those people. The cycle is simple: Big things are suggested, grand visions, immediately upon which that small minority with loud voices begin telling us why we shouldn't.

Why we can't.

Oh, they're so clever.

Mars is a great example. Expect the resistance to this to heat up as it gets closer to reality, mainly in the form of very studied reasons why we can't achieve this next grand target. Why we can't live there, why this, why that, rarely with any proposed solutions.

Simply why we can't.

Watch for it.

It's already begun. The "can't do" crowd sees the writing on the wall; they're already trying to get ahead of it. Their squawking will get louder and more floppy-armed, apoplectic-desperate as we get closer, as more targets and milestones are reached.

Count on it.

The more real Mars gets the more squeaky the noises from the can't-do crowd will get.

Some of us will choose to listen to them.

Those of us the history books will write about, however, will be the ones that choose to ignore them.

The history books will write about the ones that figure out a way.

"But ..." one might be tempted to think. "It really is ..."

Look.

We're gonna do it.

We're going.

No matter how many degrees you have, no matter how smart you may think you are, no matter how much you may think you know better ...

You don't.

So stop wasting everyone's time (including your own), cleverly explaining why we "can't", and start helping the rest of us figure out how we can.

Man on Mars is inevitable. As is Man among the stars. The question is only, and will always only be, how much resistance the doers among us have to fight through.

Getting there will already take *huge* ingenuity, resources and sacrifice, even if *everyone* on the planet was behind it.

It ain't gonna be even a *little* bit easy.

So pitch in, would you?

The challenges by themselves are great enough; we can do without the addition of the wholly unnecessary challenge of fighting against those who think we should give up.

Come.

Be a doer.

Be one who finds solutions.

Copyright David G McDaniel & Michael C Petry

If you catch yourself being a can't-do person, consider changing your tune. Instead of:

"We can't do it because XYZ."

Try:

"To do it we need to consider XYZ. Here are some ideas for how we might solve that problem."

That's really all it takes.

For any worthwhile objective (and space is quite worthwhile), the sequence should be:

1. We can't do X because of Y.
2. Here are some ways we could solve Y.

Those two steps go well together.

The more pointless chatter and resistance the rest of us are forced to endure, the more people shout "Can't!" without offering up solutions, the longer until those next cool things become reality.

But they *will* become reality.

Audacity

IF WE WANT A THING (wanting it being the key, of course), we *will* have that thing.

Wanting it, to be clear, means wanting it with no question, 100%.

When our certainty hits that level, things get done.

The cycle is something like this:

1. Name it. (With full clarity.)
2. Want it. (With full intention.)
3. Get it. (By taking the necessary action.)

The bigger those demands, the greater the action necessary, and thus the greater the progress.

Audacity: a willingness to take bold risks.

To get anywhere we'll need that in abundance.

We'll need to be ready to ignore the small minority that has only negative things to say. Reasons why we shouldn't. Why we can't. Why it's hard. Why it's dangerous, why it will never work and on and on.

Please. Go away and let the rest of us get the show on the road.

What's that saying?

Lead, follow, or get the hell out of the way.

Sorry, but negative people frustrate us. Greatly.

Yes, we have the audacity to think big.

So sorry that you don't.

Those people do eventually "get there", of course, but only in the wake of and on the coattails of the rest of us, the ones who faced up to the challenges and solved the big problems. The ones who didn't shrink from life and decide things couldn't be done.

Fortunately the world is filled with people that *do* get things done. People taking action. Each of us have opinion leaders we respect, heroes, those we look to when framing our own actions.

That impulse scales.

Groups, nations—the entire world can point to individuals at each level they look to, pushing the boundaries of progress. People they count on; people paving the way. People who have high aspirations for humanity. People who are those audacious thinkers.

These individuals not only create our new world, they inspire us to do more. If we're not directly pushing those boundaries, it's our job to encourage and support the ones who are.

We all win when we do.

Forty Suns is meant less for those progress leaders already taking action, more for the rest of us. The idea being that we begin thinking like them. If we each dreamed of and took action like an Anousheh Ansari, for example (to use but one of a great many specific examples)—no matter where we started; a dollar in our pocket or a million in the bank—imagine what a difference that would make.

The very tapestry of the world would transform.

That's the power of the can-do spirit.

Good news is, all it takes is a decision.

Each of us can decide to be part of the can-do crowd.

Naming A North Star

WITHIN THE LIFESPAN OF YOUR AUTHORS we've gone from dialing telephones (yes, they had dials) hardwired into the wall (no, they weren't mobile), to cell phones (no dials, no buttons, no wires), to say nothing of the other technologies that have continued to roll forward like gangbusters—along with exponential increases in our understanding of the world and the universe around us.

The world is moving fast.

Which excites us greatly.

What's a good North Star for that progress?

A North Star, for any enterprise, person or movement, is a metric that captures a long-term vision; one that's always there, provides direction, inspires, is clear and visible and, most importantly, is attainable by striving in its direction. By that definition we see humanity's North Star as a sort of aggregate measure, summed into one overall thing on which we can focus that defines our evolution:

Power.

Power is what will get us there.

Much of our success depends on power. How abundantly can we power our lives? How fast can we travel? A good (if somewhat dark) fictional example of what our logical, methodical progression into space might look like is the TV series, *The Expanse*. That's a good representation of how we might eventually move around our own solar system. The more famous *Star Trek* gives us an idea what things might look like in a future with gobs of power, and the ability to travel really fast. Gobs of power is a good goal. Having that kind of power on tap will open up the stars.

And so Power, we would say (see Chapter III, including the various Revolutions that have marked our progress as a planet), defines us. Power is the hallmark of our world.

Power, how sustainably we make it, how easily, how much is available to each of us, is our North Star.

Character

UNDERPINNING THAT MEASURE OF PROGRESS IS, of course, us. Power is easy to measure, us ... not so much.

How to quantify the real North Star of our space future?

How to measure us?

Character, the ability to rise, is within each individual. Great explorer of the unknown, Ernest Shackleton (who has a lunar crater named after him, btw), posted this when looking for crew for one of his history-making Antarctic expeditions:

"Men Wanted: For hazardous journey. Small wages, bitter cold, long months of complete darkness, constant danger, safe return doubtful. Honor and recognition in case of success."

There's that Frontier spirit that has made our world what it is today. These days we have better amenities in our various space programs than Shackleton was offering, but the general idea remains. Going into the unknown, that sense of derring-do, risking things no one has done before, demands a certain character.

Shackleton also had this to say:

"Difficulties are just things to overcome, after all."

Taking things overly seriously, apparently, may not be the best approach. See to everything you can see to, of course, but at the end of the day our attitude in how we attack the obstacles before us can have just as much impact on our success.

After all, it's only a game, right?

Doing Our Part

WE DON'T ALL HAVE TO GO TO SPACE. We don't all have to invent new rockets, or discover new and fantastic physics.

In fact most of us won't.

But by being interested, by keeping up with what's happening in the fields of space, by sharing with others, by being a fan, we help that progress more than we may realize.

If we are, each of us, actively interested, at whatever level we're playing the game, progress becomes possible.

In fact it's all but guaranteed.

CHAPTER XVI
Turning Point: Space

"Success is not final, failure is not fatal: it is the courage to continue that counts."

— Winston S. Churchill

MOST OF LIFE, it's said, is simply showing up. Whether mentally or physically, just arriving is half the battle.
We think this is true.
The world we live in was crafted by us, for us, and we did it by showing up and doing stuff. It's hard to show up and not at least do *something*. Thus, the fact of simply being here is perhaps the best start possible.
Thanks for showing up.
Next is to see how far we can take it.

The Upward Spiral
THERE'S A THING CALLED A VIRTUOUS CYCLE. It's like spinning a flywheel, adding momentum every time you give it a shove in the direction you want it to go.
The official definition:

"A chain of events in which one desirable occurrence leads to another, which further promotes the first occurrence and so on, resulting in a continuous process of improvement."

Space Age 2.0 has begun, which means the flywheel of progress is spinning. Companies and agencies are actively pursuing opportunities in orbit and offworld, nations are making it part of their agenda, each act of which adds momentum to the system.

Imagine one of those playground merry-go-rounds.

You might be a big kid, with both hands on the bars, shoving like a mad-man to make it go faster and faster. You might be a smaller kid, throwing on a hand every few times around, adding as much to the fun as you can.

Surely you remember those days.

You've probably also been one of the multitude of kids standing on the fringes, watching in rapt enthusiasm, cheering and laughing, encouraging the main players to give it their all. Or even just an approving parent, farther across the playground, on a bench with your coffee, smiling silently at what a blast all those crazy kids are having.

Support is meaningful.

All contributions are valuable.

No matter your role, the result is energy and enthusiasm added to the system. An upward spiral of success.

That and a whole lot of fun.

Thinking Extraplanetary

How do we begin thinking in these terms?

As with most decisions, it starts about as simply as possible:

By making up your mind.

Once you do that, and the next person does, and the next, and they tell two friends, and they tell two friends ...

Like the flywheel analogy, as more and more of us begin to participate, as more of us pay attention, as more of us make even a passing awareness of space events part of our daily lives, things start to happen.

Extraplanetery means living on other planets. *That* reality is further off, but it's where we're headed, and it will one day (perhaps even sooner than we think) be a part of our everyday existence.

To get there our minds have to be there. Even if only a little. Achieving goals means thinking of them as if they were a foregone conclusion. That works both individually and as a group. If we decide we're an extraplanetary civilization, then the things we do will tend to align in that direction.

Our focus becomes our reality.

How should we frame that vision?

Unless we skip a few steps of scientific discovery and end up with warp drive way sooner than expected, Mars will very likely be the next actual planet we occupy. In order to stake that next claim, to see that extraplanetary thinking through to reality, that is where our global focus will most effectively lie.

The stages might go something like this, putting energy into each in sequence and in concert:

1. Building and expanding our on-orbit presence.
2. Making the moon an every-day reality.
3. Mars.

Mars, the Red Planet, will be the first big milestone on the road to our larger space future.

Another Way To Frame It:
Why is Mars important?

A thousand reasons, some more obvious than others, but there's one that might not seem obvious at all.

Call it framing your target.

As a younger man, one of your humble authors trained extensively in martial arts (Kung Fu specifically, because, duh, Bruce Lee). A lesson taught then had to do with punching *through* the target—the idea being to leave all the energy of the strike where it needed to be.

Humanity can (and should) frame a similar concept when it comes to the moon and Mars.

Last time we went for the moon we made it, barely, then kind of faded. The moon became the short-term endgame.

To make it stick we need to punch farther.

If we truly want to make the moon a no-doubt, for-real, actual everyday reality, we need to punch *through* it.

By punching through the moon, aiming for Mars, the moon itself becomes a foregone conclusion.

Once we're established on the moon, the rest of the solar system opens up nicely.

Mindset

THE FUTURE WILL NOT JUST HAPPEN.

Okay, the future *will* happen, one way or the other, but wouldn't we rather be more in control? If we expect the future we want, we must reach for it and grab it. We must create it.

What might the future hold?

That is entirely up to us to decide.

As we said in our opening remarks, way back at the beginning of this book, it's a great time to be alive.

All these pages later and we still believe it.

—

There are a million things to be done, a million ways to improve, all of which could be distilled down to two things we humans must do if we're to secure our future:

1. We must be better to each other.
2. We must create options to this one-world existence.

Both challenges are huge. We've had opportunities to work on #1 our entire existence. We've never had an opportunity to legitimately work on #2 until now.

Perhaps we might find solutions to one in the other.

Yet, even if we can't quite get a handle on the first, we can't wait any longer for that to be perfected. Sure, if we don't take care of our personal issues we only drag them with us into space, but at least by expanding, by giving ourselves more places to fight, we at the same time give ourselves more places to live. More places to work on our manners.

It would be a shame to come this far and simply fight ourselves into oblivion right where we started. Or get wiped out by a natural disaster because we never figured out how to escape this world if we needed to.

The Money In Space

HAVE WE MENTIONED MONEY?

Of course we have.

Let's talk a bit more about moolah. More specifically profit potential, which is what will drive ongoing interest in space.

Fortunately opportunities to make money in space are being worked on every day, by enterprising forces around the planet.

Just wait till we find gold on the moon. Moon bases will start popping up like popcorn, with rocket launches going off like fireworks.

The Space Rush will have officially begun.

The inverse of the money to be made, of course, is the money that first needs to be spent.

Money is what it will take to get there.

Lots of money.

Good *god* is space expensive.

New technologies = expensive $$$
Advanced research = expensive $$$
Program development = expensive $$$
Launch and operations = expensive $$$
Space exploration = expensive $$$

And on and on.

Money represents energy. How much energy are we willing to commit to achieving a desired end? The conquest of space will take tons of energy. Is it worth it to secure that future freedom?

Throughout time, when we decided something was worth it, we found ways to make it affordable. Computers are a great example. There are plenty of examples. Pretty much everything in our history is an example. Building the first big ships that could sail around the world took a *ton* of resources and manpower.

Risk.

Courage.

Was it worth it?

Yeah. So we did it.

Same with constructing canals, building the first steel skyscrapers, digging the first undergrounds ...

You get the idea.

Don't throw money away, of course.

Watch the budget. Of course.

But it's like buying groceries. You don't *not* buy groceries. Maybe you don't get the most expensive cheese, but you get cheese. You eat, no question. Groceries are a given part of the household, and not just rice and beans; you throw a few snacks in there too. Things that make you happy. Things that lift the spirit.

So should it be with space.

Space should become a necessity.

Following that logic, space, it occurs to us, must become the groceries of humanity.

(Just when you thought we were getting good at this.)

An Example

AND SO COMPUTERS. You've probably heard how meager was the computing power aboard the Apollo moon landers back in the 70s. Those missions had 4 KB of memory, with 32 KB hard drives. (KB is a Kilobyte, or a thousand 'bytes' of digital data, a term now rarely used.) Today we typically use either MB (Megabyte, a million bytes) or GB (Gigabyte, a billion bytes).

Lately we've even been breaking out the TB (Terabyte, a trillion bytes). We love our big-number prefixes.

When it comes to the systems used on Apollo, for comparison an average phone these days has something on the order of 32 GB of storage memory. Eight million times that of the lunar landers.

Cost-wise, engineers got the Apollo computer module 'down' to $15,000 in the end. Meaning back then the memory in your phone would've been worth, strictly by the numbers, something on the order of $120B.

Your cell pricing plan doesn't sound too bad now, does it?

That's to say nothing of the chip technology, the display tech, the cameras, the software, the (most especially) internet connectivity, and all else that would've made your phone a state secret and a national treasure back in 1970.

Point is, computers became important. We found value in them, we wanted more of them, so we kept throwing money at them to make them better and better and, critically, more affordable. Eventually getting them to the point where they are today.

Computers are the foundation of our entire modern society.

Satellites, interestingly, are the thing most driving space innovation right now. Why? Once again we're back to the demand for our personal devices and conveniences. Our demand for faster better more-data more-coverage communication connections (that very internet connectivity) for our computers, devices (phones, TVs, etc), has driven a veritable arms race in the satellite—and therefore rocket launch—arena.

This is good.

We'll take it.

It demonstrates how demand drives results.

Because of this we've continued to develop more and more cost-effective launch systems to get those satellites into orbit. Which means the cost per kilogram to launch a payload into space has been dropping like a mic at a rap battle.

Here's a sampling of what launch costs have looked like over the last four decades:

Space Shuttle (1981)	$ 85,000/kg
Minotaur-C (1994)	$ 35,000/kg
Space Shuttle (1995)	$ 27,000/kg
Delta II (1997)	$ 19,000/kg
Falcon 1 (2006)	$ 10,000/kg
Atlas V (2016)	$ 6,000/kg
Falcon 9 (2017)	$ 2,000/kg
Falcon Heavy (2023)	$ 950/kg

Pretty sweet example of that cost vs demand thing.

And the drop is accelerating. SpaceX's Starship will bring cost down to just $100. NASA's goal for 2040 is "tens of dollars per kilogram".

Dang.

But it's just Rocket Economics 101.

Economies of scale.

Another virtuous cycle; more launches lead to lower costs, which lead to cheaper payloads, which lead to more launches and so it goes. The more rockets pounding launch pads around the globe, the faster we send more and more payloads into space, the higher the launch cadence, the better we do.

Always more.

Never less.

Why would we ever consider slowing down?

If we run out of science missions and satellites, start building the refueling stations we'll need in orbit, get our next-gen space stations underway, fire up the first zero-gee manufacturing facilities, begin putting things up there we know we'll need for our moon adventures and on and on. Affordability and sustainability go hand in hand; we get there by building an ecosystem that demands it. Launch providers and payloads, in conjunction with destinations such as space stations, depots, lunar goals and access to other infrastructure.

Space is far too monumental, far too difficult, far too challenging to let slip even an erg of momentum. With anything as impossible as space, easing up, deciding maybe it's a little too tough now that we think about it, that maybe we've done enough for now, that we should take a step back and figure out a few other things first ... that sort of thinking, when it comes to something as hard as space, risks the whole enterprise sliding backwards.

Not on our watch.

And so we say More.

Never slowing, always faster.

Like Billy Idol grunting his *Rebel Yell*, More, more, more!

(Is our passion bleeding through?)

That's the only way we become extraplanetary.

Demand

WE ALLOCATE RESOURCES according to what we deem important. Time and again, throughout history, we've shown we *will* pay for—and get—the things we want. From the personal all the way up to the level of nations.

We humans are resourceful.

Satellites are but one arena. Many advancements will define our comprehensive offworld presence. Many more activities will be needed. Many, *many* more things to spend money on.

Space is expensive.

Question is, will we decide it's worth it?

Random Note:

By 2030 the luxury travel industry—just luxe, not regular trips to Disney and such—is forecast to hit $2.3 trillion, a good chunk of that going to accommodations. One luxury suite in Dubai currently runs $100,000 *a night.*

Crazy?

Apparently not; supply can't keep up with demand. True. Places like that can't become available fast enough for people to come spend their money. As we noted earlier, the commercial space sector is projected to reach a total value of $1.4 trillion by that same date: 2030. Meaning the lovely people of Earth are expected to be willing to allocate a cool *trillion dollars more* each year to spending short stretches of time in ultra-swanky digs than they're prepared to flow toward advancing our space future.

Fall on whatever side of that debate you will; for us it seems an imbalance of interest.

And that's just luxury travel. What about luxury things? Three of the largest companies *in the world* today are luxury brands. (LVMH, Dior, Hermes, to name them.) Luxury firms in general were more profitable in 2022 than American tech firms.

Yeah.

These are companies that sell things like fashion, perfumes, cosmetics, jewelry ... you know, fancy stuff. We're apparently willing to pay handsomely for things like that.

Which is great, we're not knocking it, we (your authors) like fancy stuff too, but these examples do illustrate a point.

If we'll pay thousands for a purse, or a night in an Earth-based hotel, surely we can spare a few bucks, as a civilization, to make sure that—one day—we're able to flash that pricy purse at a lunar hotel. Or on a space vacation.

Right?

Otherwise it will just have to keep being us in some fancy place here on Earth.

A castle.

A yacht.

Dubai.

Same old same old.

Forever Earth.
How passé.

Space Vacations?

SPEAKING OF SPENDING MONEY, what *about* space vacations?
Space for the rest of us.
That's the goal, right?
A life in space?
At least for those who want it.

Like most new things early forays by us civilians have been, and will be, expensive and, therefore (mostly, so far), only undertaken by the very rich. The people willing to forego that extra Bugatti or a few of those Dior purses for a rocket ride. That said, even at this early stage there have been exceptions; people who got selected to go to space though they were not Scrooge McDuck. In those cases someone else footed the bill.

But it's vital that they did. And do. It's important that these rich people jump in and spend big money so that, eventually, space is doable for everyone. If no one was willing to fork out big money for early versions of (name your tech), that tech would never have taken off and ... that's right, never would've become available for the rest of us.

"Early adopters" are vital. Pioneers, who commit and risk either money or life or both, have funded the advances that have gotten us where we are today.

And so as those prices come down (and they will), sending more and more private citizens affordably (and relatively safely) into space will become the next rung on the ladder toward a life for *all* of us in space, should we choose to go.

Which is where our true space future begins.

Until very recently, most of what we thought about when we thought of space was centered around career astronauts, military test pilots, scientists and professional-others (brave teachers, trained researchers and so on); those intrepid adventurers who manned the ramparts of the next frontier. We watched, and we cheered, while they went.

That's changing.

Astronaut Jessica Meir put it perfectly:

"We're bringing the spirit of the entire planet with us."

Those pioneers, like Jessica, are making it real for the rest of us. Once the average person has the option to live on the moon, for example—should they have a moon residence on their bucket list—in our humble opinion that's the mile-marker for the dawn of the *next* space age.

We'll name it now:

Space Age 3.0.

Everyday people going to and coming from space. Living on or visiting the moon. Space stations.

Will that happen in our lifetime?

Seems likely if we keep up this pace.

Until then the travel of private citizens to orbit will continue to evolve, expanding in parallel with everything else that's happening in the pursuit of space. That future is already underway with the advent of so-called "space tourism". Let's call it space adventurism for now. Sure, it's more of a novelty at the moment, but that's okay. Many great things started that way.

Next will come actual vacations.

PUSHING THE ENVELOPE

If all goes according to plan, starting next year Space Perspective, a company out of Florida, will begin taking people to the very edge of space—rising above 99% of the Earth's atmosphere.

And they'll do it in a balloon.

Called *Spaceship Neptune*, this extraordinary balloon and cabin was developed to be the most accessible, most sustainable, and safest spacecraft on or above Planet Earth. Over the course of a six-hour cruise (twice the length of the one the passengers in *Gilligan's Island* set sail on), riders will ascend in the comfort of a cabin with nine reclining, plush seats, with spectacular—nay, spine-tingling—views from full, 360° wraparound windows.

Plus, the cabin can be reconfigured to accommodate a special event, like a dinner for two, or a wedding.

What?

Yes.

A wedding in space.

Who will be the first private citizens to do that?

—

So what might a space vacation look like?

Hotels in space? Orbital designs have existed for some time.

Hotels on the moon?

Imagine a lunar vacation. Seeing the footprints of our first astronauts to walk there. Other landmarks. Descending magnificent craters, stark in their beauty, your $5,000 Dior purse slung smartly across your shoulder.

(Apologies, we had to make that joke.)

But how cool would it be? Having fun in low gravity; all the things you could do. Look up at the moon from Earth and you see Tycho Crater. How awesome to be able to point to that, wherever you are in the world, and say:

"Yep. I've been there."

The more we make the moon part of our everyday reality, the more secure our future.

Lava tubes are thought to spider-web pretty much everywhere beneath the moon's surface. What if we turn a few into subways? Then you could be whisked around quickly and easily to all sorts of destinations. Lunar circumnavigation, anyone?

Sounds trippy (badump bump).

So-called "moon culture" will set the stage for how we adapt to and live on new worlds. Maybe we'll even call those who go "moon people", the next set of pioneers that solve the issues the rest of us will face when extending our lives beyond Earth. In the same way the pioneers of the West blazed trails and learned the ropes for living successfully out yonder, so too will the moon people.

Making it easier for the rest of us to follow.

The beauty of the moon is that it's not far. We can experience all sorts of things there, conduct experiments, try different techniques and more, practice our extraplanetary existence and do so with the confidence of being not too far from home. Getting back to the environments we're used to will be a snap.

So many reasons we need to be on the moon. Soon.

(Nice rhyme, eh?)

Space: The Great Unifier

EVERY ASTRONAUT WHO'S BEEN TO ORBIT says the same thing.

It's life altering.

To a man, to a woman, they have the same experience. A oneness with their fellow humans.

Sounds like a great epiphany to have.

In fact, it's been suggested (by those people who've been) that going to space should be a prerequisite for any leader. Before a new official assumes office, they should have to go up there and have a look down here.

To see.

To get that perspective.

The experience even has a name, the "Overview Effect". For each person who experiences it the Effect drives renewed interest in maintaining our world.

As more of us go up there, out there; as more of us experience this effect; well …

We can only imagine the results.

For one, it will drive ever faster that flywheel of interest.

And it just may make us appreciate our fellows, and this planet of ours, that much more.

—

Pioneer space company, Stoke Space, has a relevant quote:

"We've reached a critical point in our history ... How do we grow as a civilization without destroying our home? We believe a perspective from space will give us the information we need to achieve both."

We couldn't agree more.

Group Power

AS NOTED IN THE GOVERNANCE CHAPTER, we tend to group ourselves into ever more complex, ever more significant ruling bodies.

If five or ten kids in a schoolyard agree so-and-so is the Grand Poobah, then that person is the Grand Poobah. That might last for a recess period—or maybe less if the kid elected is mean and no one wants to play anymore—but the Poobah gets that honor because everyone decided and agreed. (Yes, Poobahs can become Poobahs through coercion, or by beating up other kids; the idea we're going for here is that anyone in charge is there because we, collectively, agreed—whether through votes or submission—to let them be.)

Get a few hundred million people involved, over the course of centuries, playing a game with a long history of agreements as to how and when the elections of Grand Poobahs are done, what they mean, along with powers-of and hereby-granteds, duly-appointeds and all else, and it gets a lot harder to simply stop playing the game and go to lunch.

In most cases we're kind of stuck with these vast and many-peopled governments we've agreed to.

One advantage of these big groups, however, is significant.

Resources.

Nation States are the backbone of space. They have the resources. They can (and will) spend money when everyone rallies to the cause, such as in the space race of the late 1960s. They field the expensive scientific programs that lead to advancements; they have the potential to bring together the focus of many.

Yet that backbone is only part of the group equation.

As we saw in the first space race, national pride and determination can (and did) move mountains, but without an ongoing objective by which private industry—and even individuals—could profit, things withered.

Perhaps, even without industry, a concept like *Forty Suns* back in the 70s could've kept public demand high and therefore governments would've added to space budgets rather than cutting.

Maybe.

It's more likely that, even with public demand, if there was no money to be made in space those programs would've eventually scaled back. Governments are not run for profit, not anymore, which is why most modern nations are swimming in debt.

All the pieces need to work in concert.

Scaling from the largest we have:

1. National interest and investment
2. Private enterprise and profit potential
3. Group focus at every level
4. Individual passion and support

We're getting it right this time. Much more private industry, and even individuals, are seeing the value in space, and this collective cooperation—and purpose—is the fuel that will drive us past the inertia that might otherwise hold us down.

Space Age 2.0 doesn't have to be another almost-ran. It's in full swing, right now, and ramping up.

Let's take advantage of that state of affairs.

That's true group power.

If we, each of us, do our part to increase the momentum, to spin the flywheel, the net result will be progress.

Strong Individuals = Strong Groups = Positive Change

Science Fiction

THE MILITARY-INDUSTRIAL COMPLEX, and industry in general, drives much of our technological progress.

Artists, in turn, are the dreamers of our civilization. That much has always been true. It's a symbiotic relationship, if at times uneasy, but it is the relationship with which we've always worked. Proper balance must be struck.

The dreamers and the doers, together, will get us there.

Of late some have even taken to inserting an "A" for "Arts" in the STEM acronym, making it STEAM. An insertion that isn't off the mark. The artists of the world, as noted, share their own, vital role in our advancement.

A Global Effect

WHEN IT COMES TO THE DREAMERS of our collective hope, never in history have the aspirations of so many been focused so far into the future. Pop culture, movies, books, comics, music, art, games and all else, have been planting the seeds of a science fiction reality in our imagination for decades. We are, as a global culture, primed more than we've ever been to look ahead and imagine fantastic things. Science fiction has already influenced and inspired scientists, engineers, teachers and other professionals around the world to join the space field, to envision incredible advances and more. That Alcubierre drive we mentioned back in Chapter II?

Mister Alcubierre was inspired for his idea by the warp drive technology in *Star Trek*.

And remember the United States Space Force logo?

Star Trek through and through.

So many examples.

As a planet we tend to think in sci-fi terms. Whether we've seen *Star Wars* or not, whether we went and watched *Avengers* (along with 100 million other people opening weekend), in this modern age that tapestry of the fantastic forms the backdrop of our thoughts, views and expectations. Our fictions define us. Increasingly we'll see those fictions become our reality, as well as bearing witness as they inspire the thinking of each new generation.

We're reminded of a soon-to-be-famous quote:

"Today we build the artifacts of our sci-fi future."

We can't recall where we heard it, but pretty deep, no?

In our sci-fi future we'll look back on the things we accomplish today in the same way we look back on the pyramids: With a sense of awe, and, most importantly (we believe), with a deep appreciation for the dreams and visions of their creators—and the sheer amount of effort and willpower those past generations were willing to (and actually did) expend in the pursuit of their ambitions.

Those generations being us.

Space Opera vs Fantasy

ONE COUNTER-POINT.

Before we lay it out, though, and head toward a wrap on this chapter, let us say: thank you for listening. We've clearly had much to relay on these various topics, and we know we can at times (often?) sound like a broken record.

We do appreciate your understanding.

Our final point for this chapter has to do with the subject of Space Opera itself. Space Opera is the name we might give the reality of a truly extraplanetary existence. Living on other worlds. Traveling—eventually—among the stars.

Until now we've been pitching hard for just such a state of affairs. To bring our space future into being. Yet we realize such a future is not a guarantee of teacups and rainbows. Possibly, even, worse things will come with the good.

Cyberpunk lunar gangs?

Martian warlords?

There's a beauty to simplicity. To a life on the farm, say. Or a fantasy setting; castles and princesses and noble quests. Even more wholesome ways of life; something akin to an *Avatar*-like backdrop, for instance. Existing in harmony with an energizing ecosystem. There are ways to have such an existence right now, right here on Earth.

Yet that existence will live within a world of technology.

One can, for example, with effort, get off the grid. Lead a simpler life, if one so desires. Again, what we're pushing for is more playing fields, not a way of life.

Space Opera may be cool, we certainly love it, but for some of us it may just be a necessary evil in order to safeguard our future.

Yet it will be required.

Some of us may not want spaceships and laser guns. Some of us may prefer dragons and magic.

Dragons and lasers?

In all cases Space Opera *is* the next step. If we're to have those other playing fields we *must* make it to space. If we don't want to risk everything we know evaporating in a one-world catastrophe, we need to create those options.

Adventure is out there. Opportunity.

In all its forms.

But we have to get there.

We have to create it.

So let's bite the bullet and make it happen.

The simple fact of the matter is, Earth can't be our only place to play.

By the Numbers

Here's what's happening in space:

- 72 countries around the world have space programs, 16 of which have launch capabilities. The most notable being NASA, but there are also agencies in Russia, China, Europe, Canada, Japan, and India.
- As of 2021 some 10,000 private enterprises worldwide, large and small, were engaged in some form of space tech.
- 4 crewed launch capsules are in service globally, with 6 planned, not including the heavy lift vehicle, Starship.
- Over 100 launch companies around the world have launch vehicles in various stages of development and use, with SpaceX, RocketLab and ULA (United Launch Alliance) being the premiere private players.
- There are 450+ private satellite companies and providers active.
- 3 lunar missions are queued up for the coming years, in the US alone, after which routine trips will begin. The crewed Artemis 2 launch, the Artemis 3 crewed lunar landing, the Artemis 4 docking with the Lunar Gateway, with future yearly landings on the Moon thereafter.
- 3 rovers are currently in operation on Mars as of this writing: Curiosity, Perseverance and Zhurong.
- More than 250 robotic spacecraft and probes have ventured into space since we first began exploring beyond Earth's atmosphere in 1958.
- 2 space probes, Voyager 1 & 2, launched in 1977, have traveled into interstellar space. Voyager 1 has traveled the

farthest distance. Of the probes we've launched into space, five will end up leaving the Solar System: Pioneer 10 & 11, Voyager 1 & 2, and New Horizons.

So, yeah. Space Age 2.0 is here.

CHAPTER XVII
T-Minus Zero

"Occasionally, or often, life is more about seeing what you can get into, rather than what you can get out of."

— A wise person

FORTY SUNS INVITES US TO TAKE A MOMENT to snap out of our heads. To adopt the view that our current conflicts and dramas seem challenging only because we've agreed that they are. To decide therefore that, if we really wanted, we could play a much bigger game. One where the vast majority of us win.

To decide we can embrace a new frontier.
Something far more grand.
Rewarding.
Fun.
Sound doable?
On the one hand we're talking a simple decision.
Decisions being where all great things begin.
On the other, we're suggesting quite a seismic shift in how most of us currently view our world.
As the countdown reaches zero we say the time for such a collective decision is now. We happen to be living at a pivotal moment in history. (Congrats for making it, by the way.) A point at

which such a grand decision could, actually, perhaps for the first time, be made into a reality.

And we may just have the makings of a solution right here.

Forty Suns could become the "killer app" for this new frontier.

Not an app, per-se (though one day soon, maybe), but you get the concept. Apps solve the things we need solved. Apps generally make life easy and convenient. *Forty Suns*, in all its forms, including future, actual apps, exists to make the reality of our space future that much more likely. With *Forty Suns* as inspiration, the lot of us will (hopefully) actually keep our focus on that future, solve the things that need to be solved and, ultimately, end up with lives that are easier, more secure, and more convenient.

And way cooler.

Think of *Forty Suns* as an ion engine, versus a nuclear rocket. Steady pressure, your interest in space and fandom driving things forward, a continual build of speed, each impulse building upon the last until … soon we're racing toward our objective:

An extraplanetary life.

Let us again point out the fact that there are so many smart people in the word, so many of them putting that intelligence to use, doing smart things, coming up with smart solutions … it's nearly preordained that we *will* advance.

Those smart people, throughout the ages, have gotten us where we are today. The latest generations of smart people, those alive with us now, are getting us where we *will* be.

They don't have to go it alone.

If we're not one of those super-smart people, creating the big pieces of our space future, we can actively support them. We can be the fuel for that long-term ion engine.

Which is what *Forty Suns*, as we keep repeating (there's that broken record), is all about. Steady pressure; a light focus from all of us.

Paused here at T-minus zero we feel confident.

We humans have a great history of finding the right solutions in the end.

—

May we suggest a Trifecta of Truth? To go along with and support the above. Three very workable guideposts by which to center your journey.

1. Live your best life.
2. Help others.
3. Space is the answer.

Taken as a decision, each creates a more prosperous reality.

We've found challenges resolve more easily when framed within those very simple, very powerful truths.

Level Up

How can you level up?

Your "Level" when it comes to Space Age 2.0 is really a state of mind more than anything. However, exercising our remarkable foresight and legendary ingenuity (thank you, thank you), we came up with a few ways you could gauge your progress.

We imagine five basic levels:

1. Space Lord
2. Space Captain
3. Space Marine
4. Space Fan
5. Space Newb

The following are a handful of samples for each, of what might constitute actions to be taken in order to be considered a bona fide member of that level. These are only suggestions. Each sample could spin off a whole category of its own.

Our purpose in any of this is merely to inspire.

Space Lord

A TOP-LEVEL SPACE DIFFERENCE-MAKER.

A few ways we might classify such a personality ...

- Gone to space.
- Made a discovery that leads to new space science.
- Founded a space-based company.
- Heads a national space interest.

Space Captain

A SPACE ACTION-TAKER AND ACTIVIST.

We'd consider this anyone who is ...

- Building rockets.
- Doing space science.
- Widely delivering space news and interests.
- Teaching space and science.

(*We realize, as we're creating this list, the fact of writing a space themed book—this one—kind of makes us Space Captains. We swear it wasn't planned. We're not worthy!*)

Space Marine

ANYONE ACTIVE in the pursuit of our space future.

Such as ...

- Working at a space company or national space interest.
- Forming space-themed groups.

- Actively attending conventions; investing in space.

Space Fan

THE CORE OF OUR SPACE FUTURE.
   The heart of Space Age 2.0 lives here …

- Follows the space news, sometimes avidly.
- Trades space news stories.
- Mentions space stuff to friends and family.

Fans are the at the core of any successful thing.
Fans are the bedrock.
Fans:

- Drive loyalty
- Create culture and community
- Make more fans
- Bring attention to their obsession
- Generate word-of-mouth

In short, fans are the centerpiece of any movement.

—

Famed rock band AC/DC knew how important were their fans. They even had an anthem, specifically saluting the fans who made rock-and-roll the juggernaut that it is:
For Those About To Rock.
On that original tour, in one of the most epic concert experiences ever (possibly *the* most epic concert experience ever), they brought actual cannons, mounted them above the stage—then fired them in a 21-gun salute.
The scene went something like this.
The band, on stage, shouting the line:

"For those about to rock ... !"

Dramatic pause.

"Fire!"

BOOM!

"We-sa-lute-you!"

Paired with a sharp, emotional salute. An ode to the fans who made their existence possible. A salute for the fans of rock everywhere.
AC/DC got it.
As do we. In the same way the importance of fans wasn't lost on them, neither is it lost on us. Being a fan means everything.
So for that ...
We salute you.

Space Newb

ANYONE JUST GETTING STARTED. If you're reading this book, especially all the way up to here, but are not yet engaged in any of the above, then you are, at a minimum, a newb.

Congratulations.

Whether you level up the ranks or not, you are at least aware of the huge amount of progress being made toward our space future.

We, and all and everyone at every level of the space community, are terribly glad to have you.

—

As with other sections of this book, we're open to input. Other ways to categorize—and thus inspire and encourage—are welcome.

Formula For Success

WHEN IT COMES TO MAKING CHANGE, a formula applies:

Awareness + Interest = Change

Whether change you make directly, or change others make as a result of that interest (the true mark of a fan; any activity succeeds through the driving force of the interest of its fans), successful forward motion depends, to a greater or lesser degree, on the elements of this formula.

You're reading *Forty Suns*. Your awareness has been raised. Your interest, we hope, has been stoked.

Change for the better will be the inevitable result.

Our People

THIS IS OUR NEXT BIG CHANCE to see this through. To start taking command of our future in space.

Ready?

It's an exciting time, and we happen to be alive right when it's happening. Future us's will look back and thank those of us alive today for what we did, for how we stood up and made the effort to secure that space future.

In many ways it won't be easy.

At times it may feel like a fight.

Turn the negative to a positive.

Tell the naysayers to keep it down.

It's too easy to stay on the sidelines and find fault, which means many will do it. (How fun, after all, to sit and hurl scorn?) This includes huge news organizations, of course, but, believe it or not, even—at times—supposedly "scientific" publications. Our would-be allies, too, fall into the trap. When it comes time to decide between reporting on the brilliant ideas being posed to overcome tough challenges, or instead doing a piece on cost overruns, they often choose the latter. Giving stories that juicy, pessimistic, negative twist is just too lucrative.

It's in our nature. Even our so-called friends of science are susceptible.

Thus it's up to us to fight that propensity. Both in ourselves and in others. Solve problems that need solving, don't harp on them. Don't read the negative stories. Cancel subscriptions. Demand positive reporting. Refuse to hear ill of the people and programs attempting this herculean task.

Those are the very people and programs that need our support. Not our barbs.

Head in the sand?

Nope.

Our heads are high above the mountains.

After all, that's the clearest view of this island we call Earth.

Get 'Er Done!

An old American Southernism, "Get 'er done!" evokes an attitude worth adopting. Let's call it "making things go right", or the tendency to accomplish tasks and create a winning future.

This attitude wouldn't necessarily represent a static state. It would more be a state of action, as in working toward a betterment of self, things, others, the world. When it comes to taking action there are four general types of people:

1. Those working to make things go wrong, or actively preventing things from going right.
2. Those happy to watch things go wrong while offering no solutions.
3. Those complaining how wrong things are and not liking it, but doing little or nothing to make things go right.
4. Those making things go right.

There are more Type 3s among us than any other, and that's a product of many factors. It's easy to complain.

We need more Type 4s.

A Few Thought Exercises

HUMBLY TENDERED FOR YOUR CONSIDERATION. A few things here at the end to make you go "Hmmm".

—

The World Is Your Oyster

AS A THOUGHT EXERCISE, imagine the world is yours. The whole damn thing. Everything and everyone in it.
But it's the only one you have.
There's currently no way to get another one if this one goes kaplooey. Some of the games you play on this world are kind of risky. Some could even mess it up, bad. That happens and ... no more world. All the fun stuff you used to do (assuming you like to do fun stuff) goes away. Your world, no more.
How would you guard against that?
Sure you could be more careful, and maybe you should, but accidents happen. One big enough "oops" and that's it.
If there was a way to insure against that loss, a way to have other places to play (as needed), would you do it?
Even if the effort was high?
Would a guard against extinction be worth your attention?
Would it be worth working for?

Too Much Help?

ANOTHER. Extrapolate what we've been pitching here in *Forty Suns* and imagine, if you will, what the outer limits of such a concept might be.
What if all 8 billion of us suddenly started thinking about space?
If the entire world, every man, woman and child, suddenly had it in mind either to know about what was happening in the space industry or, perhaps worse, to actively get involved ...
Would that even work?
As hard as we're pushing in that direction, that sort of "logical end" might yield gigantic traffic jams. Are there enough projects on the lineup? Right now we humans are launching a bunch of satellites, planning a few new space stations and upgrades,

working on how to establish a presence on the moon and, ultimately, Mars, only ...

How much help can any of those take?

Seems like they need *tons* of help, and they do, but would a sudden, total, global enthusiasm overwhelm rather than assist?

In order to picture that scenario we could imagine it like a barn raising.

Say you and ol' Zeb are out trying to hoist up a brand new barn, with all the fixin's, and you're desperately looking for help. You expect a few of the neighbors to show up, hoping maybe they'll bring a few friends to pitch in. That'd be perfect.

So you put out the word, like we're doing here with *Forty Suns*, and you say, "Hey! This thing we're doing is a great thing and it will be perfect for the community, and if we only had more barns how great would that be? Come on over and help!"

Your goal is to get that barn up, maybe even another. You've got enough wood for at least three, actually, plus you have plans for a silo—if you can get to it.

Suddenly, all of New York City shows up.

9 million eager people, interested and ready to help.

Um ...

What happens in that scenario?

You'll get your barns up, that's for sure, probably in record time, and your silo, but damn if that's not a lot of wasted potential.

For this thought exercise, then, try to imagine how our current space industry could possibly respond if *everyone* on the planet suddenly wanted in.

Where would we start? What would everyone do?

How would we maximize that desire to help?

Life Outside

IMAGINE YOU LIVE ON A BED.

Fun.

You've never left it.

Maybe not so fun.

You can see the bedroom. You can see a few things out the window. From what you see out the window—other houses—along with evidence you can perceive from the bed, you conclude the house your bedroom is in may also have other rooms. Rooms you could get to.

Only thing is, you've not yet figured out how to get far from the bed. In fact, you've only figured out how to walk around its edges. Kind of exciting, to be off the bed, inspiring, even, to see it from that angle (Look! I can see my pillow from here!), but you always need to keep a hand on it. That's as far as you've been. The rest of the room looks quite intriguing. You've figured out how to throw things across it. Even though *you* can't go anywhere in the room, you can see those places (a desk on the far wall is particularly fascinating), and you can toss things that reach those places. You've landed a few socks on the desk.

Even more exciting, those possibilities only have to do with the bedroom. How great would it be to be able to go check out the rest of the house? Let alone the stuff *outside the window*. Those other houses out there. Who knows who might be living in them. And who knows what might be beyond even them.

The world outside the window looks like a really vast place.

You see the analogy.

Being stuck to Earth is like being stuck to the bed.

Laying in bed is nice. In fact, truth is the bed is our refuge. We were probably born in a bed, we'll probably die in one. We have our best dreams in bed.

But never being able to leave the bed? Never being able to venture out into the world?

No bueno.

Yet that's the way we've always lived.

Really imagine such an existence. Imagine seeing all that out there, all around you, but not being able to get to it. To touch it. To experience it. Imagine being stuck to your bed. Being left only to look, and to wonder, and to dream.

We don't know about you, but that makes us a little sad.

Luckily, for the first time, at this moment in our existence, we can actually begin to figure out ways to venture across the room. How to get up and go sit at that desk. After that ... maybe even how to go to the rest of the house.

And outside?

One glorious day, we might even make it there.
But only if we make it our priority.
The bed is comfortable.
The bed is easy.
We can't stay in the bed forever.

Despicable Me

ONE LAST THOUGHT EXERCISE. We're pretty much full of them, so apologies.

As we shoot out into the light of a brand new tomorrow, here at the end of this long, inspiring (hopefully), *Forty Suns* tunnel, let's make this last one a dark one.

Just for fun.

Assume a secret cabal *does* exist. A shadow group dedicated to ... anything despicable. Pick your evil cause, and these guys are up to it.

Even *they* should be behind a concept like *Forty Suns*. If they're smart enough to run the world from the shadows, surely they *can't* be idiot enough not to realize that same world that's fragile for the rest of us is fragile for them too. Right?

We die, they die.

Pretty dumb if they're thinking otherwise.

Hello? You're stuck on the same rock as the rest of us.

Yeah, yeah, we know. You're king of the biggest international drug slash terror slash arms-dealing black-lotus triad in known history. Hollywood writes action movies based on your epic bad-assness. You eat lesser villains for breakfast, dining on their livers with some fava beans and a nice chianti. You're dark. You're to be feared. You haunt the most epic tech-noir dance clubs, where the beautiful people vogue in thumping clouds of smoke and blacklight lasers (lasers!). You know way more than we do. Obviously. You understand how the world *really* works. You make deals in your Doctor Doom fortress of dread we could never comprehend ...

Blah, blah, blah.

You're stuck here just like the rest of us.

Get over yourself for a minute.

Join the human race, at least for a bit, and start playing a *real* game. Help with the stuff that matters. We're actually fun to work with, believe it or not.

Would it help if we said we have cookies?

Come over from the dark side. Put your shoulder to the winning wheel—at least until we're not tied to this one planet—and let's see what we can do. Help us get the show on the road.

Then, when we're done, once we have a few more options ... if you haven't grown to like us by then, feel free to go back to being Bond villains.

Better yet, *Space* Villains.

A whole new era of evil.

See?

Everyone improves in this game.

Come on, despicable guys. Help us get there.

Expanding The Wins

THERE'S A LOT OF ABLE-BODIED PEOPLE looking for work. Whether a crisis of leadership, a crisis of the free market, or a symptom of the times, this failing can be addressed. An accelerated push for space has the potential to create many new jobs, while solving the rut in which some of us get stuck.

Not being able to work is as much a morale problem as it is a financial one.

Earning a buck is great, and vital, but giving able people a chance to work directly toward a rapidly expanding future would not only be a move toward survival it would be a boost to morale.

For everyone.

These aren't high-minded ideals.

One Last Sports Analogy

AT A FOOTBALL GAME 20+ PEOPLE are on the field playing. (We happen to like football, so that's been our go-to.) Another 100+ or so are on the sidelines. At least a few hundred are covering the game.

80,000+ are in the stands. Millions more are watching on TV. Cheering.

Each is key.

The game doesn't happen, at least not meaningfully, without each of those groups. Not each of us has to be on the field, in the newsroom or on the sidelines. Most of us will be home or at the bar watching and cheering. But that's perfect.

Because at the end of the day we, the fans, are what make the whole thing go.

To extend the analogy, when it comes to space we're on the goal line. One yard from the end zone. As a team we've come all the way down the field, from hunting and gathering our food to building grand cathedrals to here, now, eating nachos while we watch the next great rocket launch.

We're so close. Closer than we've ever been.

Let's come together and punch it in.

Let's not get a penalty, lose yards, or, worse, turn the ball over.

If you're a fan, cheer. Stand in your seat. If you're on the field, play hard. If you're a ref, let them play ball. If you're a decision maker, call the plays that will help us win.

Be bold.

Take the shot.

The time is now.

If we don't make it into the end zone on this play, we may have to put in another long drive to get back into this position again.

The Struggle Is Real

LET'S ADMIT IT. We humans have been through a lot.

Hardship, struggle, great loss, often just to survive.

But let's also admit we've advanced. Let's truthfully acknowledge we're closer to the conquest of space than we've ever been. Hardship, struggle, loss—these have been part of every grand venture, every human enterprise worth undertaking, as they will be part of this next great wave of success.

Let's prepare ourselves for that.

Relish it, even, if we can, knowing from our own histories that such struggle is what leads to the win.

Above all let's support each other as we each do our part, as we each endure our own struggles, to ensure a vast, wide-open future, not only for ourselves, but for the many generations to come.

How?

HOW DOES THE SIMPLE ACT of us "paying attention" to space matter? How does it make a difference?

Where does any of this lead?

We're glad you ask.

(Don't you love the way we set ourselves up? By the way, we're nearing the end.)

It would, at first blush—even at second and third, and perhaps even fourth—seem that, with something as terrifically huge and overwhelming as space, just making note of stuff other people are doing would have little to no effect.

How can us paying attention move the needle?

Directly, it probably doesn't.

Several big steps must be taken in order for us humans to push our space reality past the point of no return. The point past which we're unlikely to lose ground.

In general order those steps would be:

1. Solidify our on-orbit reality, with occupied commercial stations, manufacturing, refueling, orbital asset manipulation and management, debris oversight.
2. Establish an orbital presence at the moon, infrastructure for control of lunar and even system operations,

communications, future mission staging, deep-system launch-points.
3. Complete moon bases, begin resource collection, manufacturing, exploration, experimentation, trans-lunar transport, mobility, civilian presence.
4. Mars.
5. Asteroid mining and the larger moons.

After that we could be reasonably certain that, short of full civilizational collapse, we're in space to stay.

From there, of course, we continue further into the system and, eventually, beyond. Those first five steps must be well established first.

How do each of us play into that?

Our attention and interest have power.

Each event in the pursuit of our space future, each milestone, does not happen in a vacuum. (Okay, it's space, so maybe it *does* happen in a vacuum, but you get it.) Participation scales. It's never an all-or-nothing proposition. The people delivering the final product, the astronauts, the engineers and so on, participate at the highest level. They, in turn, depend on those directly supporting their efforts, mid-level participants such as those who provide funds (be they private or state), which in turn depend on demand, which in turn depends on ...

You.

Our individual participation may appear to be practically nil, in the overall, but the mere fact that we demonstrate that demand, that we're following events, talking about them, encouraging the curiosity of others ... our attention, though it may appear slight, *does* create an effect.

Where we put our focus matters.

Interest is vital to advance.

Demand drives progress.

And so "how" we do it starts with turning our eyes to space.

Each of us. All of us.
Everything scales from there.

Adieu Ado

HERE WE SAY GOODBYE TO ALL THE FUSS.

Adieu is French for goodbye. Ado is Late Middle English for "a state of agitation or bother, especially about something unimportant". As in "much ado about nothing".

By saying "adieu ado" we're kissing our troubles goodbye.

And why not?

The future looks bright.

Let's push that positivity far and wide.

You've probably heard of trickle-down economics. This is the theory that, in essence, good things that happen at the top of the food chain will eventually trickle down and benefit everyone.

We propose an inversion of that principle.

Call it trickle-up positivity. Better yet, trickle-up purpose. The idea being that change starts with us. The free-thinking individuals that make up our society. Sometimes called grass roots, paying it forward, even karma, it's true that the betterment of our world begins right here.

Yes, it's difficult not to be the effect of the massive machine over our heads, pushing down, but that machine is made up of, guess what? Individuals. Other "us"s. Which means if each of "us" is putting positive effects into our environment, if each of us is talking about, encouraging and participating in the things that will build our space future, those effects then push outward, in all directions, creating a better world.

Advances are happening all around. Each and every day the scientists, governments and corporate enterprises of the world are pushing the envelope of research and development, bringing to reality the technologies that will get us there.

Forty Suns, with any luck, has made you aware of these things. With any luck we've made you want to know more.

Regardless of the motivation, the important thing is that progress is being made.

So ... what now?

We wrote a whole appendix on that, Appendix A.

In the meantime all we ask is that you tell a friend.

In fact we encourage you to spread the word. Give a copy of *Forty Suns*, or share yours, or tell others to buy their own. Not so we can make a few bucks (though that will be one result), but to spread this forward-looking point of view.

Or just give them your copy when you're done.

—

Bringing us to the real kicker:

What did you learn?

Forty Suns didn't teach you how rockets work. Or the theories behind faster-than-light travel. We didn't even really cover much about the exploration of space. Our goals have been more broad.

No, our primary goal has been to shine a bright light on *everything* that's happening in humanity's pursuit of the next frontier. To invite you to look. The hope being that at least a few of you will go forth and expand your own interest, and knowledge, perhaps even become active players, and that the rest will become advocates of our space future.

We need as many of us as possible with our eyes—at least part of the time—out *there*. People who are unwilling to let things slip. Who make the decision that they're living in this time, in this *now*, with conditions as they are, with opportunities as they exist, and that they can *do* something about it.

Are you with us?

In the Foreword we said *Forty Suns* wouldn't teach you all there is to know about space, but it *would* arm you with curiosity.

We hope it's done that.

Curiosity may be the most powerful weapon we have.

So to the question "what did you learn?" ... If anything you've probably learned just enough to be dangerous. Dangerous to the threat of ignorance. Armed with what you've learned it will be hard for ignorance to cast its sleepy web over you. As well, you've become aware enough to be dangerous to the possibility of Space Age 2.0 being forgotten, or put on a shelf.

That one most of all.

Revisiting that sports comparison (for, like, the fifth time; apologies), for every season ticket-holding fan who paints their face and stands in the seat yelling at every home game, there are hundreds who simply wear a team shirt or tune in for the scores.

Fans at every level are what make sports-entertainment the juggernaut that it is.

The same will be true for our space future.

Whether you simply check on a few key launch events now and again, or actively follow all developments in science and space, sharing that info with anyone who will listen, your participation—at any level—is what will keep the fire lit.

The Bullets

To summarize the various concepts we've been pounding the drum for, all we would suggest is that you:

- Be curious.
- Take responsibility.
- Focus on the positive.
- Remember the vast majority of your fellow humans—in their hearts—are well intentioned. Misguided? Maybe. Bad people? No.
- Spread the word that our future is in space.
- Take to heart the reality that this is our next, great opportunity to pull off the first major milestones on our journey to the stars.
- Make it your mission to promote the successes in our space industry.
- Celebrate and support your fellow fans of Space Age 2.0.

In short, let's play a bigger game.
And let's do it together.

Has Anything Shifted?

HERE WE ARE AT THE END. Let's take a final look at that system pressure.

Worst case for us, your authors, would be that your outlook for our space future pretty much stayed the same. Actually, worst *worst* case would be that your confidence dropped. Shame on us if it did. Boy did we mess up.

In fact, worst worst *worst* case: what if you were at the bottom when you set out on this journey, and lost even more faith? Does that mean we turned you into an *enemy* of space?

Yikes.

That would be really bad.

But of course we're hoping you saw a glimmer of possibility at the outset, and we've actually managed to raise your outlook.

Alright. Time for a little introspection. Now that you're here, when it comes to space, would you say you:

1. Don't think there's any chance of us ever being able to make a home anywhere but here, on Earth.
2. Think we might one day live on Mars or something, but not for a real long time.
3. See possibilities, but aren't sure if we humans can get out of our own way.
4. Are hopeful, with a pinch of optimism.
5. Believe not only can we can make space our next frontier, we will.
6. See that progress in space is happening already, and would encourage all involved—plus the rest of us—to pour the coals on the fire.

Our hunch, just to say it, is that most of you started around 3, and landed here at the end near 4.

Which, if true, would be awesome. If more of us were hopeful with a pinch of optimism, that shift alone could mean huge things for our progress in space.

Be honest, of course. This is between you and yourself (me, myself and I?), so no need to pretend. Now that you're here, do you feel any more positive? Has anything shifted?

Our hope is that you do, and that it has.

Destination Space

IN A WAY, GETTING TO SPACE IS OLD NEWS. Been there, done that.

From a wider view, and to use yet another analogy, we've really only put a raft in the pond behind our house and demonstrated we can float.

In that analogy there's a whole world out there, so much of it accessible by water.

Lots and lots of water.

Vast stretches, seas and oceans, much of it treacherous.

Now that we can float ...

What opportunities does that present?

When it comes to space, we've conquered the pond.

The question then becomes, where to next?

To that we say, *Ad Astra.*

(In case you forgot, that means "to the stars".)

About Your Authors

WHAT ABOUT US?

We, Dave and Mike, are perhaps the least qualified to scribe such a book. That very lack of qualifications, however, may be what makes us the *most* qualified.

After all, this is a book for the uninitiated. One intended to relate to readers who may have little or no awareness of what's happening in the space industry—our goal being to draw attention to the reality of Space Age 2.0. Who better than a couple of guys who are super interested in that very thing? Two dudes who are finding out about most of it themselves?

We're only glad you read all the way to the end.

Seriously, thank you.

May we make a suggestion?

Have a cigar, maybe a Scotch, or maybe a hot hoagie and a beer, if you're more in the mood for that, and ponder what might be your own role in our space future.

APPENDIX A
What Now?

SO YOU'RE LIVING IN THE OPENING ACT OF SPACE AGE 2.0.
Super congrats for that. Seriously.
Where do you go from here?
By reading this book you've already taken a pretty big step. *Forty Suns* is the antidote for debilitating FOMO (Fear Of Missing Out) when it comes to our space future.
Only, now that you've plowed through these factoids, opinions, insights, zany and not-so-zany ideas ...
How do you keep staying space-smart?
All those words you just read are fixed, and won't be any different until the next *Edition* comes out.
Lots will happen in the meantime.
Most of it pretty damn cool.
At least a bit of it pretty damn important.
Lucky for you, we have more ways to stay engaged.

—

You now have an idea of the scope of what's happening out there. An inkling of just how broad are the world's space interests.
In order to keep up, we believe a few lists are in order.
That said, lists can be boring.
They can also be super useful.

And so here we've put together what we hope are a few super-useful ones, ones you'll like, ones you'll use, broken into categories of resources for staying up on:

1. The Cosmos
2. General Science
3. Rockets & Launches
4. Space Agencies & Programs
5. Space Commerce
6. Ways To Get Involved
7. Ways To Become Informed
8. Ways To Help

Note, these lists are being regularly updated on our site:

FortySuns.com

Let's dive in.

The Cosmos

DISCOVERY, the things that keep us informed as to how our universe works, what's out there, etc.

There are many fascinating things happening among the stars, with lots of people covering it.

Dr. Becky
YouTube Channel: @DrBecky

PBS Space Time
YouTube Channel: @pbsspacetime

SEA
YouTube Channel: @sea_space

Astrum
YouTube Channel: @astrumspace

General Science

THE STUFF THAT UNDERPINS EVERYTHING ELSE.
Without science, we don't advance.

Kurzgesagt
YouTube Channel: @kurzgesagt

Learning Curve
YouTube Channel: @LearningCurveScience

Fermilab
YouTube Channel: @fermilab

Big Think
bigthink.com

Brilliant
brilliant.org

What Da Math (Anton Petrov)
YouTube Channel: @whatdamath

Sabine Hossenfelder
YouTube Channel: @SabineHossenfelder

Rockets & Launches

THE THINGS THAT GET US THERE, and the science that drives them. Seems they should have their own category.
So we're giving them one.

Everyday Astronaut
YouTube Channel: @EverydayAstronaut

Space Launch Schedule
spacelaunchschedule.com

Rocket Launch Live

rocketlaunch.live

Spaceflight Now
spaceflightnow.com

Space Agencies & Programs
THE NATIONAL GROUPS MAKING IT HAPPEN.
Support your local Nation State.

NASA
nasa.gov

ESA (European Space Agency)
esa.int

JAXA (Japanese Aerospace Exploration Agency)
global.jaxa.jp

SSA (Saudi Space Agency)
ssa.gov.sa/en/home/

Some of them offer cool launch kits you can grab, too.

Space Commerce
COMPANIES AND ENTERPRISES GETTING IT DONE.
Many are rocket companies, so as you explore you'll find overlap, but these are great places to start your discovery.

SpaceX
spacex.com

Varda Space Industries
varda.com

Axiom Space
axiomspace.com

Rocket Lab
rocketlabusa.com

Hermeus
www.hermeus.com

Skyrora
skyrora.com

United Launch Alliance
ulalaunch.com

Firefly Aerospace
fireflyspace.com

Stoke Space
www.stokespace.com

Ways To Get Involved

HOW CAN YOU TAKE PART?
Perhaps the best list of all, right here.

The Planetary Society
planetary.org

National Association Of Rocketry
nar.org

Space Force Association
ussfa.org

NASA Solve
nasa.gov/solve/

Students For The Exploration & Development Of Space
seds.org

Space Generation Advisory Council
spacegeneration.org

Aurelia Institute
aurcliainstitute.org

Limitless Space Institute
limitlessspace.org

Breakthrough Initiatives
breakthroughinitiatives.org

Earthlight Foundation
earthlightfoundation.org

National Space Society
space.nss.org

Ways To Become Informed

STAYING UP-TO-DATE, becoming informed ... this is how we keep pace with the fast-moving evolution of our space future.
Consider subscribing to or following one or more of these.

Payload Newsletter
payloadspace.com

Space
space.com

Space News
spacenews.com

Tech Crunch
techcrunch.com

Interesting Engineering
interestingengineering.com

Phys.Org
phys.org/space-news

Ars Technica
arstechnica.com/space/

Ways To Help
AT THE END OF THE DAY, this is what it's all about. How can we help make it a better future for all?

Rigel Institute
rigelinstitute.com

STEM Advantage
stemadvantage.org

Space Foundation
spacefoundation.org

Space Center Houston
spacecenter.org

Space For Humanity
spaceforhumanity.org

—

These are samplings of what you can find and who you can follow in order to stay up on the latest. As you may imagine, there are plenty out there. By beginning to explore the limited resources given above (in those super-cool lists), you'll find yet more sources that match your interests.

One thing leads to another, as sung so convincingly by 80s rock band The Fixx.

Yes, even here at the bitter end we're quoting rock bands.

These lists are regularly updated on our site, at:

FortySuns.com/Explore

Note, when it comes to YouTube channels, and any of the sources given above, bear in mind some are more positive in their outlook than others. Some are afflicted by the impulse to play upon the train-wreck syndrome from which we humans so gravely suffer (and respond so beautifully to). Some prefer to draw eyeballs with alarming headlines.

All, however, are dedicated to reporting the news of technology, science and space. As with the regular news it's up to us to find the real stories. All are attempting to highlight the path(s) to our future in ways we can understand.

No matter their tone (and most are upbeat), their hearts are in the right place.

They, like us, see the value in chasing our space future.

—

Last caveat.

As you watch videos and explore sites it's easy to run down rabbit holes. *Lots* of eager content creators out there, which is awesome, but as you navigate the intriguing labyrinth that is the internet, trust your Spidey-Sense and check what you can for validity and accuracy along the way.

Find sources you can trust.

It's easy to get lost among channels and sources that maybe don't know what they're talking about. As has been famously said, everyone is entitled to their own opinion, but not their own facts. Opinions are fine. Facts are facts. In many ways this whole book is one big, fat opinion. But facts matter.

On your journeys through life and adventure, confirm the facts where you can.

Don't get caught way far afield, having absorbed a whole bunch of "knowledge" that turns out to be little more than someone else's hopeful delusion in the end.

Our space future depends on the facts.

—

How else?

What else can you do to stay informed and be a part?

Investing. Putting your money where your mind is. Interested in that cool new invisibility-shield tech? Find the companies that are advancing it and, if they're public, buy a few stocks. Or if there are other ways to help, do so.

Truth is, just keeping up can be overwhelming. There's *so* much happening around the world, so many places to land our attention, where do you even start?

Finding focus helps. Pick a favorite subject in the field. Become a grassroots expert on X, Y or Z.

Lots of ways to get going.

Maybe you're looking for a new purpose. Or maybe you're looking for a purpose, period. Space Age 2.0 could definitely be a worthy, and worthwhile, cause to throw yourself into.

Or ... maybe you're having trouble getting motivated.

First off, if that's true, we didn't do our job. If we didn't at least light a tiny spark after this many pages ... well, we've failed.

But worry not. There are plenty of motivational and self-help programs and movements out there to help kick off a new you, super-charged with fresh motivation and passion.

Maybe it's time to look for your next book to read.

Last comment.

For those of you who *are* motivated, or at least want to be, and are just looking for a place to start, where your authors make their home there's a building sporting a huge, artistically rendered mural, with a simple message:

Begin Anywhere

Sounds like a great bit of advice.

—

With that we leave you a final, and admittedly quite random, "Did You Know?"

Did You Know?
Another way to stay informed.
> The British have an informal slang, "boffin".
> A boffin is someone engaged in scientific or technical research.
> The very sort of person we could get behind.
> Want to be connected? Support your local boffins.
> Who knows? Maybe, just maybe, you could become one yourself.
> You big boffin, you.

—

Don't keep it a secret that you're a fan of space. You might be surprised who else is.

Live your best life. Make your best contributions to this world. As you do, simply keep in the back of your mind that things are happening to advance our space future. Take note once in a while. Make mention here or there. Maybe wear a *Forty Suns* hat or drink from a *Forty Suns* mug on occasion (shameless plug).

Those things alone, if most of us are doing them, will remind others that we're in this together.

Thus laying the foundation for a brave new world.

APPENDIX B
Foundation

SPEAKING OF FOUNDATIONS, we want to start one.

That sounds fancier than it probably is.

It's an impulse driven by the same thing that's driven us to do all this furious typing. The impulse to make sure Space Age 2.0 doesn't fizzle. For our part, we want to do more. (Not more typing, god help us; more to advance the pursuit of space.) As your authors and founders of the *Forty Suns* concept we have, of course, written this book, that's one thing, but we also have other things in mind.

Lots of things, actually.

As we go to press we're piecing together as many of those ideas as possible, ways for people to get involved, ways to support those who already are.

First will be developing our site.

Seems an obvious place to start.

We can almost hear the website headline now, yelled by a proper carnival barker:

"*Thee* home for fans of Space Age 2.0!"

There we'll gather as many great resources as we can, with links to cool stuff. A portal, a hub—*FortySuns.com* will become the (thee!) gathering place for these ideas.

And hopefully a few of you.

We'd love to have you join us for the latest and greatest.

Next would be to eventually get some groups going. We're not sure how or in what form, or pretty much anything else, but the idea of more of us out there talking with each other and keeping awareness high is appealing. Stay tuned to the site. We'll put things there, plus you can contact us with your own ideas through the site as well.

Forty Suns

How would such a Foundation work?

For starters it would, ideally, look far ahead, to its namesake.

Forty Suns.

The idea of a truly exraplanetary existence for humanity.

Meaning, in truth, the logical conclusion:

Extrasolar.

That's a big prize. At this early stage it's also a huge reach.

Not, probably, within the grasp of this generation, or even the next (though stranger things have happened).

That said, while we work on the things we can accomplish now, our ultimate focus should be on that distant horizon.

The required technological progression of which is crazy enough to warrant our grandest analogy yet.

Ready?

We do love our analogies.

Thee Analogy

Forming a Forty Suns Foundation now, in the 2020s, dedicated to reaching the stars, is like forming (would've been like forming) a group in the 1500s called the Lunar Society.

A society dedicated to putting colonies on the moon.

In 1500 that level of futuristic science and tech would've been a similarly long-range view.

A literal moonshot.

Still, such a society could have existed back then. Dedicated to one, major mission: encourage and celebrate the scientists, inventors and engineers of the day. Those taking bold steps, making bold proposals. A society dedicated to bringing broad attention to scientific achievements in the near-term, creating an environment where discovery and invention could thrive.

Leading to advancement after celebrated advancement.

Leading, one day, to the first colony on the moon.

Could a Lunar Society founded in the 1500s have gotten us to the moon faster?

Quite probably.

Which makes us believe the time for Forty Suns, the time to begin directing our focus toward long-range goals, is now.

The result will be steady, continued advancement of life here on Earth, into the solar system and, eventually, beyond.

Ambitious?

Trust us, we're keeping it real.

Like that Lunar Society from the 1500s, we'll mostly be celebrating the achievements of the day.

In 1569, for example, it would've been spreading the news about Gerard Mercator and his nifty new Mercator map projection, which promised to revolutionize global navigation.

Wider awareness of such things back then would have moved society forward at an overall faster pace.

In the same way, a Forty Suns Foundation today will celebrate things like the latest advances in fusion, new propulsion designs—new developments in space and space-tech of all kinds which are getting us there. Not only the successes of the scientific teams making those breakthroughs, but the engineering teams bringing them to life, along with the skilled individuals putting them to practical use.

Along with all else furthering that future.

In that way we maintain our collective focus toward the next frontier:

Us in space.

Who Dares

ONE OF THE MOST FAMOUS SLOGANS in the world comes from Britain's SAS (Special Air Forces). It goes:

Who dares, wins.

Poetic, pointed, powerful.
There's a bit more. The full slogan more or less restates the first part, but fleshes out the meaning.
The full slogan is:

Who dares, wins. Who sweats, wins. Who plans, wins.

We've made this slogan our own.
Will you join us?
Stay tuned.

APPENDIX C
What Might The Future Take?

WE USED TO LIVE IN CAVES and fear tigers in the bushes. We no longer live in caves and no longer fear tigers in the bushes. Today we live in cities and fear other things.

One day we'll live on entire other worlds and have new fears, but the fact is we'll be *living on other worlds* and all our troubles of the present will, in the same way, seem quaint.

Progress is part of our existence.

The Moonshot Mindset

SO WHAT ARE WE DEALING IN?

Growth. Big Think. Gains. Pushing the technology envelope. Creating bold futures on all fronts.

In short, the "Moonshot Mindset".

It's a term we've used in this book. You've probably heard it before, coined during the Space Race of the 60s, where it epitomized the thinking behind the literal *slew* of incredible, impossible, can't-be-done items on the checklist to get a man on the moon.

No way was it doable.

But we did it. It's that sort of collective will that makes things happen. The Moonshot Mindset very much encapsulates the purpose of *Forty Suns*, in that we mean to get the world thinking of things not only as possible, but as welcome challenges. A large group of people did it once, it can certainly be done again. The result was the introduction of a seemingly impossible thing into our everyday lives. The more a thing is done, the less it costs, the fewer the surprises. Travel by jetliner is a perfect example of the progression of a moonshot mindset to an everyday thing. The idea of that was, at one time, equally out-there.

Now it's quite routine.

A reminder for each of us. Say we ourselves never intend to fly. That hardly means planes aren't needed. Our future demands routine spaceflight in the same way our current civilization demands planes. Whether we ever go to space or not.

In that sense *Forty Suns* is more of a rallying point than anything else. A place for us to come together over a common objective:
Space.

If someone is talking *Forty Suns* they're talking progress. They're imagining a future they deem possible, and they're out to support those who will invent and create the means to achieve it.

If they're talking *Forty Suns*, they share this conviction:

"Space is the answer."

If you're here, making your way through these final (and often wordy) appendices, we assume you believe that too.

We assume you share the Moonshot Mindset.

Responsibility

ANYONE AS OLD AS US, your authors, can likely recall those Chief Seattle campaigns back in the 70s. For a time they were pretty much everywhere. Bumper stickers, billboards, TV commercials. The wizened Native American chief, shedding a silent tear, with the caption that the earth does not belong to us, we belong to the earth.

It would seem, however, the opposite is true.

The Earth is our responsibility.

Circling back to the potentially volatile nature of the world and our own fragility, let us point out that we have a huge responsibility to ourselves and to our future selves to take charge of this planet and manage it responsibly. The Earth *does* belong to us, and as soon as we forget that, as soon as we get careless with that possession, like anything ...

Stuff breaks.

When it comes to us humans, when you bring us into contact with something an expected progression of mastery can be observed. First we adapt, then, inevitably, we control. In 1854, when that Chief Seattle quote was first made, man's mastery of his environment was crude and generally very localized. We're now much further along, and though the world is far from being within our full control, each year, each decade, each century we gain more understanding, developing more ways to mold our environment— even as we safeguard and improve conditions.

It's true we've been mis-controlled so often at so many points in history by so many bad actors that "control" is often seen as a bad thing.

It isn't.

Good control is how we manage our environment, which in turn is how we advance our living standards and way of life.

Good control is not destructive. With good control all factors advance.

Responsibility at each step is key.

—

Another responsibility we share as imaginers of the future is the responsibility to ask questions. Constant, inquisitive seeking of answers. Think of the most basic question:

Why?

Where might asking Why? lead? "Why" is a great question for just about anything. Think of any reality and ask, Why? "Why did humans choose to invent pants?" When you have that answer, ask again Why? to the answer.

Keep asking Why?

Might you reach an end to that road? Might you run out of Whys? What *is* the ultimate answer?

Fascinating to think of that eventual, final truth.

The average kid asks 300 questions a day (that's a scientific guesstimate). If we, as adults, did that it would get annoying fast, but you get the idea. Curiosity has gotten us killed (along with many cats, presumably), but it has also gotten us where we are today. Without curiosity, yes, many of us would've lived longer than we did, but those of us around today would still be eating berries from trees and fearing those tigers in the bushes.

All hail the curious. We owe them.

So be responsible, and be curious.

—

Let's put that in a short list:

- Take responsibility for your immediate environment, your world, your future.
- Ask questions (and get answers).

Ethics

WE'VE MENTIONED taking responsibility, we've told you to live and have fun, even while reminding you your existence is fragile ...

Okay if we give you one more piece of the framework?

Forty Suns supports ethics in the application of science and technology. Ethics apply to each of us. Proper ethics are seen to by each individual, always with consideration toward benefiting the greater good.

Any of those Bond villains out there listening?

This game, Space Age 2.0, is about making things better for all of us. War, exclusivity, and other applications of technology, as noted, are typically the drivers of innovation, and yet they don't have to be. Rockets were missiles first, planes got used for war before we started flying them to grandma's for Christmas—a lot of our best technologies have evolved that way. War and conquest will always be part of human expansion. We're not saying that's all bad. Profit in particular, the quest of, generates loads of positive advantages. All we're saying is that at *Forty Suns* we're out to embrace a bigger game, with a much bigger purpose.

Ethics being its foundation.

We might say there are two related, yet distinct, thrusts for the development of space. Overlaps exist, as always, but the two main drivers of our gains are:

1. Profit motivated action (vital for growth)
2. Future motivated action (altruistic)

As such, where either is concerned, *Forty Suns* is out to support and encourage the ethical development of technologies and science that push us toward our collective space future.

Greatest Good

WHEN IT COMES TO ETHICS we've heard it said there are no solutions, only trade-offs.

That's certainly one way to look at it.

It's true there are no solutions where absolutely everyone and everything wins. Does that mean there *is* no universal solution? Only a trade-off?

Perhaps.

Thanksgiving dinner isn't great for the turkey. A summer stroll in the grass is bound to kill a few bugs. This is life. "Greatest Good" is the way we all win, sure, though sometimes that means a few of us lose. In *The Wrath of Khan* (best *Star Trek* movie), Spock gave us his version of an old truism, "Logic clearly dictates that the needs of the many outweigh the needs of the few."

Another you've probably heard, less poetic but just as true: You have to break a few eggs to make an omelet.

Pick your saying. There are many.

In each case it's up to us to decide what's most important, then do that thing.

Not everything, or everyone, can win.

But that's okay. Trade-offs may be a part of life, they don't have to bring us down. The best solutions are when we, as a group, trade up to a brighter future.

The Elephant In The Room

FORTY SUNS IS A LOOK TO OUR SPACE FUTURE.

But wait.

Shouldn't we (not the 'royal' we, but the 'we the people of this Earth' we) be concentrating on things here at home?

Of course.

That will always be true.

However, should we ever allow ourselves to become so consumed with local problems that we don't look out? That we don't dream of, and work toward, more?

Absolutely not.

We don't dispute:

- Space is hard.
- Space takes incredible resources.
- There are problems here on Earth.
- You, us—everyone needs resources too.

There's no question "doing" space takes a lot of effort.

Sacrifice.

But we must be doing it.

For all our sakes.

And so *Forty Suns* is pushing for that near-term—and mid-term, and far-off—future. The advances that will ensure we never collapse under the weight of our local problems.

Yes, someone needs to tend the farm. But at least a few of us need to be out, exploring and settling that next horizon. Finding the things that will make tending the farm easier.

This is us.

And so, though focused exclusively on positive energy—and with a purpose of inspiring the same—any of the things we talk of in this book, like all things, will find detractors. After all, it's easier to complain than to innovate.

Therefore it's worth a moment here at the end to touch on this rather silly subject.

CAVE People

IT'S INEVITABLE THAT SOMEONE, somewhere, will oppose absolutely anything. Think of the most perfect, most worthy thing, and someone will be against it. At the beginning of this Appendix we mention living in caves, bringing to mind a funny term which sums up the idea of general opposition:

CAVE people.

Or, "Citizens Against Virtually Everything".

There have been CAVE people throughout time. As an exceedingly small minority they were rarely heard from. Primary reason for their silence?

1. There was no platform for them, and
2. They were so few in number.

Their numbers have not grown.

The availability of megaphones and soap boxes *has*.

Now, with the advent of some of the technologies we've talked about in this very book, social media and broad internet access in particular, anyone and everyone is afforded a voice—including the CAVE people.

And sometimes they're the loudest.

An illusion, to be sure, but because of this relatively new opportunity to amplify their opinion these "dissenters against all things" can make it feel as if they hold more influence than they actually do. As if there's more of them than there actually are.

They don't, and there aren't.

Their voices reach an audience only to the degree we bother to listen. (So stop listening. Duh.)

Today they're called trolls. Rest assured they are not a new thing. They've always been around, and will always be.

"But," you might ask, "How could anyone be against ..."

A question with no answer, to be sure.

Someone, somewhere, will find a basket of puppies repulsive. Or at least say they do, probably just to get a reaction from others. There's no accounting for it.

So no worries. If you're here you're here, presumably, because you believe in a brighter future. Which means you most certainly are not a CAVE person. If you just happened on *Forty Suns* and are checking it out ... the odds are, again, very much against you being a CAVE person. There just aren't that many of them (see above).

In short, if you have any concern whatsoever that you might be one of these progress nullifiers, you most definitely are not.

We bring this up only to point out that, no matter how cool any of this space stuff might seem—and it *is* cool—someone, somewhere, might complain.

Feel free to ignore them.

APPENDIX D
Brave New Worlds

REMEMBER THE *MIGHTY MORPHIN POWER RANGERS?*
(Yeah, if you haven't figured it out by now, like our analogies we love our pop culture references.)
In the *Power Rangers* the lead dude, Zordon, realized that, in order to save the world, he needed to recruit teenagers with attitude.
Corny, sure, but damn did they have some wicked fighting moves. And they stood up for what was right. Zordon was wise to go for those sorts of characteristics:

- The skill and the willingness to fight for what's right.
- A good moral compass.

"Things may not always go the way in which we desire," he said. "But I believe there will be enough good people making good decisions, that good will ultimately triumph."
Also corny, but the sentiment rings true.
Stand by, because we're about to unleash yet *another* pop culture quote.
One of the great world builders of our time, Tolkien, had a knack for communicating noble sentiments through the conversations and actions of his characters. Two of his most notable, Frodo

Baggins and Samwise Gamgee, shared the following during one of the darkest moments of their journey.

Sam had this to say:

"It's like in the great stories, Mr. Frodo. The ones that really mattered. Full of darkness and danger they were. And sometimes you didn't want to know the end. Because how could the end be happy? How could the world go back to the way it was when so much bad had happened?

"But in the end, it's only a passing thing, this shadow. Even darkness must pass. A new day will come. And when the sun shines it will shine out all the clearer. Those were the stories that stayed with you. That meant something, even if you were too small to understand why.

"But I think, Mr. Frodo, I do understand. I know now. Folk in those stories had lots of chances of turning back, only they didn't. They kept going, because they were holding on to something.

"That there is some good in this world, and it's worth fighting for."

Well said, Mr. Gamgee.

Faith, we might conclude, belief in the outcomes we desire, is perhaps the key ingredient as we forge a path toward the next brave new world.

The Science Of Failure

BRINGING US TO ANOTHER great quote (one of our favorites, actually), this one by an Earthling, Winston Churchill, better summarizing those notable characteristics; the very attitude necessary to gain ground in this world. As long as there are those of us that exemplify the can-do spirit he describes (and there always will be), we have every reason to feel hope.

Churchill said, perhaps a bit more eloquently than Zordon:

"The credit belongs to the man who is actually in the arena, whose face is marred by dust and sweat and blood, who strives valiantly, who errs and comes up short again and again, because there is no effort without error or shortcoming, but who knows the great enthusiasms, the great devotions, who spends himself for a worthy cause; who, at the best, knows, in the end, the triumph of high achievement, and who, at the worst, if he fails, at least fails while daring greatly."

Which is to say, there is a certain approach that comes with any successful enterprise. Call it the "science of failure". One must be willing to "iterate", to try again and again in pursuit of the cause. To "move fast and break things", as the maxim goes.

Yes, safety is paramount.

Testing is key.

Thorough planning has saved the day more than once.

The cautious methodology is a sound one. A good one. Even a proven one.

But there's only so much that can be done before you have to actually try the new thing you're hoping to do. Some try sooner than others. The foolhardy try too soon. The over-cautious don't try soon enough. Either brings with it risks, but the intrepid are the ones that tend to fill the pages of the history books.

There's value in being willing to venture into the unknown.

Delayed Testing = Delayed Learning
Late Failures = Exaggerated Impacts (so fail fast)
Costs Increase With Time

Take a step. Put yourself out there. Go outside your comfort zone. We can all but assure you that you *will* grow. Plus be better for it. The bruises and the scars will remind you you're capable of more than you thought.

And the gains ... ah, the gains.
Those will remind you why you're here.

—

Elon Musk, probably the best iterator of our generation, had this to say when asked about the biggest challenge facing SpaceX during test flights of Starship:

"If we knew what it was, we would actually fix it before launching," he said. "So in launching, what you're doing is trying to resolve the unknowns which you cannot know before you launch, or at least we are not smart enough to know."

And that, ladies and gentlemen, is how you get stuff done.

Reality Check

A PAUSE HERE, to center our minds.
Call it a reality check.
Tagging on to our earlier CAVE people reference, there's a similar breed of animal, often called a "doomer". In their case it's not so much that they're against everything, they just don't believe in anything. Nothing positive anyway.
And so, no matter the sad cackle of any noisy doomer drifting lightly on the sunny, spring breeze; no matter any faint, discordant buzz you might happen to catch on that wonderful day you're enjoying, remember this:
Humanity has always figured out how to solve its problems.
Our ambitions have always been greater than our troubles.
We're better off now than we were. We'll be better off in the future than we are now. That cycle is inevitable.
Would we ever want to attain perfection?
Probably not.

The right balance of tension keeps us on our toes. The spirit of play, of open discussion, keeps things interesting. Do too much arguing and progress suffers, sure, but too much harmony brings its own issues. Not enough fire, not enough contrary ideas and no one innovates. We stagnate.

True Nirvana would get pretty boring pretty fast.

And so we play.

And we win.

And we lose.

But mostly we win.

Adjusting Our POV

THE UNIVERSE IS A VAST PLACE.

Not to get too hung up on that whole "huge" thing, but it's true. To imagine any significance to the human-scale events we call epic is really missing the bigger picture. In Galaxy number 87,314,013,708 of the 2,000,000,000,000 out there (in the visible universe—in what may be just one of an infinite number of universes) a super-massive star has just gone supernova in such a way that the complete annihilation of life on a dozen habited planets is all but guaranteed. Those people are way more screwed than we'll ever be. Our drama is nothing compared to theirs. Moreover, in that same galaxy thousands of other worlds won't be incinerated but certainly have troubles of their own. Catastrophe, of the sort we like to tell stories about, much of it far worse than we'll ever see.

It's a bit arrogant, then, for us to get overly serious about our issues, or to make our challenges out to be more epic than they really are. "It's the end of the world!"

Hardly.

And so operating out of panic-driven fear is not the way to get things done. (The news media and their advertisers would see this differently, but we'll cover that in the next glorious Appendix.)

Everything we do should be for our own sake, for our own ends, our own enjoyment, our own gain, our own advancement, improvement, betterment, driven out of our own curiosity, and at the end of it all our sense of accomplishment should be filled with the joy of having done the greatest good.

Action, yes.

Urgency, of course, as deemed necessary.

Panic?

No.

Desperation.

Please not.

Enthusiasm, keen interest; move fast and attack the future with gusto, not fear. Sure, you can play any game out of a fear of losing, but how much better to play to win? Both might mean you make it, but why not set your focus on winning rather than on "not losing"? The results will be the same, if not better. And you'll enjoy the ride a whole lot more.

Forty Suns considers the vastness of our existence, of who we are, and supposes we're too amazing, too powerful not to continue to take chances, make mistakes, learn and grow, expand our horizons—and have a little fun while doing it.

Through The Eyes Of Others

SPEAKING OF POINTS-OF-VIEW, ever try looking at the world through someone else's eyes? How about someone you don't agree with?

Tough, eh?

Indeed. It can be a real effort, even painful, to try and see things the way that disagreeable (to you) person does. To *really* see. To be them for a moment. To look out from their head. And even—here's the real doozie—to understand their take on things. To truly feel it for yourself, to get where they're coming from.

Why do they think the way they do?

Or, what you really want to ask:

Why they be so crazy?

Why don't they think like *I* do?

Take a look and there's a chance you'll see.

Can you imagine having their opinions?

It might just give you a grasp of their world-view.

Which doesn't mean their view is right. It doesn't mean doing things their way is the Greatest Good. But if you can see the way they do ...

That's a gift.

Arguing rarely works. Understanding will get you further.

Yes, we've invited you to ignore CAVE people, doomers and the like, but at the end of the day we're all in this together. That includes them too. Eventually it would be great to have everyone on the same page.

This is where such a concept begins.

It's a valuable exercise.

Beating The Apocalypse

DURING THE COLD WAR, when the threat of nuclear annihilation was at its peak (nuke drops could still happen, of course, but we don't obsess over it like we used to), many people spent lots of time preparing for the worst. Those early "preppers" became masters of being ready to hunker down.

Odds are fair it will never get that bad.

Though our civilization may seem fragile, and there's no doubt things could get bleak, as noted already we're pretty damn resilient, meaning the odds of an actual full-on collapse of Everything We Know (at least at our own hands) is pretty low.

Setbacks?

Absolutely. Colossal ones. Done right we could cost ourselves centuries, even millennia of progress.

Total annihilation?

Probably not.

Giant rocks may break our bones. Other cosmic events could kill us. Massive super-calderas may wipe us out. (We see you raising your hand. Basically, giant sub-surface volcanoes ready to blow. Yellowstone is an example.) But we'll probably never be able to erase all trace of ourselves on our own.

Which isn't to say we haven't shown we're capable of creating quite a lot of difficulties. We certainly have. In addition to living on a lonely rock that could abruptly, or even slowly, stop being a good place to live, we don't treat each other near as well as we should. As a solution to that, *Forty Suns* aims to give us something to look to which might bring us a little closer—while at the same time making that 'something' we're looking toward a thing that could actually save us in the end.

Space.

Seems like a win-win.

Nuclear war or a rogue asteroid will be a lot less of a threat if we have a way out. Throughout most of human existence we've at least been able to go to new lands. America is one of the best, and most recent, examples of people making a way of life *they* wanted when they didn't like what others were doing.

The days of being able to do that are pretty much gone.

Right now its Fortress Earth and that's what we've got. One world. Anything gets across the moat and we're through. We could also have a brawl in the courtyard or a fight in the castle kitchen and make things worse that way.

Never good to sit and wait. Never good to let things simmer. (See what we did there? Kitchen? Simmer? We know.)

We propose, therefore, that at least a portion of us become a different sort of prepper. Rather than hunkering down, what say we out-create any future problems? That old saying that the best defense is a good offense doesn't just apply to sports and war. By being proactive, by striking out and building such incredible competence we solve dilemmas before any real harm is done, we insure our own future. In that way we handily roll over any potential set-backs and challenges that even dare to rear their heads.

Building a fort is fine, and we may want to do that too, but if we have the ability to take action, to move at will, to seize new ground when needed, the loss of a fort or two won't matter.

We can always make more.

Think about any apocalyptic movie. Eventually they have to come out of the bunker, face the zombies and re-conquer the world.

Sheesh.

Wouldn't it have been better if they'd *really* planned ahead? If they'd been ultra proactive? What if they'd simply solved zombies before zombies ever became a problem?

We say let's keep pushing our frontiers of understanding, keep enhancing our abilities—across all fronts—until we become so capable there's no way we can come to ruin.

The Lazy & The Afraid

LET'S FOLLOW THAT UP WITH A WEIRDLY OPPOSITE TAKE. Surprised it's coming from us?

Yeah, we didn't think so.

Ever ones to take our own advice, here we're looking at this from the other point of view. The one contrary to pretty much everything we've been saying so far.

What if we were to dig real, real deep into the center of our collective psyches, only to discover we're actually ...

Lazy.

As in, we have an aversion to confronting the hard things we know we'll need to in order to succeed.

A propensity to choose the easy way.

Maybe we just want things simple after all.

You *think* an apocalypse would be harder than prosperity.

Only ... would it?

An apocalypse just means fighting your neighbors, scrounging for food—basically living like a hunter-gatherer of old.

Seems kind of chill, all things considered.

Maintaining an advanced civilization, on the other hand, is *hard*. So many demands.

Being places you have to be, doing things you have to do. So much to know about, so much to follow, so much to be aware of.

Do you suppose we might subconsciously pull back from expanding into space—an even *more* difficult civilization to maintain and forge—because it would be easier to just sit around?

Space will be exhausting, folks.

Or ... maybe we're a little afraid?

Space will also be dangerous.

Sure, we'll have a lot of neat things, shiny things, cool things, fun things, a better chance of living through the winter months and so on, but it won't be easy. And it might be loud and scary. Lots of big, fast-moving pieces.

Maybe it *is* better to just let things slide.

Go back to chillin' in the sunshine till the volcano blows.

Gamification

APOLOGIES, JUST THROWING OUT MORE COUNTER-POINTS.

It's good to look at all angles.

In none of this are we saying anything new. Sometimes, however, it's that tenth or twentieth time you hear something, or read something, that you *really* get it. It clicks. The light bulb goes on. When it does ... from that point on that realization, that knowledge or understanding is *yours*.

Like lifting a curtain of haze, that thing that was mainly just words before becomes real.

And so, though we've said nothing new, though we've unveiled no startling new revelation, we hope, this time, we've said it in such a way that it clicks. All the same games we play here, now, we can eventually play out there. We just need to get *out* there. Nothing wrong with screwing up a few playing fields if you have others to go to, don't you think? If nothing else it eases the stress.

Gamification is in so many things these days. More and more we're realizing how appropriate, how effective, how fun that approach is. Why not gamify more of life?

Why not gamify this thing called space?

How we approach our space future makes all the difference.

The Brave & The Bold

THE NEXT, AND FINAL, Appendix truly is for the brave and the bold.

Appendix E is more (ahem) motivation, if you can stand it.

If you've become a glutton for our musings, especially at this late page-count, if you *do* decide to read this final bit ...

Pour yourself a fresh cup of coffee.

Or—and this may be the actual reality (we wouldn't be surprised)—top off your drink.

You may need it.

APPENDIX E
The Gettysburg Addendum

IN 1863 ONE OF AMERICA'S most well-known presidents, Abraham Lincoln, delivered an address. In many ways, since then, that address has become synonymous with the idea of a momentous speech; the delivery of a rallying message. When someone gives a "Gettysburg Address" they're saying something they believe is meaningful, important, even key for the future.

Whether that address truly ends up being something important, a message we need/want to hear, or whether it gets used in a more ironic way, the term works. (For example, more than a few parents have probably delivered their own Gettysburg on the value of curfews.) It wasn't so much that the Gettysburg Address was long (it wasn't; only 272 words), more that it marked a significant turning point in history—while laying out worthy expectations for the future in a meaningful way.

This is our Gettysburg.

Let us say this up front: This book has been delivered in a tone of optimism, and this final (and quite lengthy) appendix comes from that same place. Though the following sections may feel a smidge heavy-handed, keep that tone of optimism in mind as you read. It may sound a bit rah-rah at times, but in the end our goal is to motivate.

And yes, we might beat a few dead horses. We hope, by the end, they deserved it.

Side note, if you haven't read the actual Gettysburg Address, it expresses a great outlook. As noted, it's short. (Much shorter than ours.) Probably take you five minutes to do a quick search and skim it.

The gist of Lincoln's address is that many great men gave their lives for a cause on that contested ground, and their sacrifice consecrated the battlefield more than the living ever could. The important thing for the living, then, according to Lincoln, was to remember that sacrifice, and carry forward the noble cause for which so many died. In other words, "carry the torch" was the best dedication that could be given, and the best thing that could be done by the living.

We feel this is true for the world of science and space.

While not battlefield heroes (in most cases), the great minds, inventors, explorers and risk-takers that gave everything they had to help build our modern age are their own form of hero.

We owe it to them, and ourselves, to carry the torch.

Here follows our Gettysburg Address.

Risk vs Reward

GREAT ACHIEVEMENTS DO AND WILL TAKE RISKS. Nothing ventured nothing gained, as the classic saying goes. You miss a hundred percent of the shots you don't take; another way of putting it. (Thanks to hockey great, Wayne Gretzky, for that.)

There are many others.

Point is, the right risk is good.

Risk has defined human advancement for millennia. Thank the gods our cavemen ancestors didn't wait until there was a cure for saber-tooth tigers before they left the cave. They took that risk. They probably had a 60% chance of stayin' alive through any given day (yes, we did just use the title of a Bee Gees song; feel free to hum it), yet they went out and *lived*. They explored. They figured out the world around them. Peril was simply part of the process.

Because of that we're here today.

Skinned knees and bloody elbows are part of the game, and worse. You, yes you, (insert name here), will one day be no more. And yes, we know we opened this book with the line "It's a great time to be alive", and it *is*.

But being alive means *living*.

Death awaits us all. Saying that while also saying it's a great time to be alive is not a contradictory sentiment. Lifetimes are finite and that isn't a grim statement it's a true one. Only once you embrace it can you truly live.

Reminding us of another old saying, something like "We're all going to die. Question is, do you go out on your knees? Or on your feet?" That's a good framing of the concept of "meaningful mortality"—making the most of our lives.

Doesn't sound like someone waiting for conditions to be just right, does it? Someone waiting for the rain to stop, or the cure for saber-tooth tigers to be figured out before they dare surge into the world and make it better.

Risk vs Reward.

Every great advance came with a degree of risk. Some higher than others. Just think of anything we take for granted today and someone, at some point, had to take a risk either to discover it or to make it known. Hell (good use of that word for this example), at one time people had to risk getting burned at the stake for spreading the word that we are not the center of the universe. Zero risk has never been a prerequisite for living. In fact, a (modern) life well-lived probably operates more at a constant 10% risk or more. Less than that and you're not really *living*.

Sound like we're on a soap box?

Maybe, a little, but this *is* our call-to-arms. Be intrepid. Be adventurous. Dare to go. Humans, successful ones, don't hide from life, they attack it. They weigh the odds, of course; there's a fine line between foolhardy and adventurous—a line that's different for everyone—but life is not lived in the dark, or in isolation.

Our point is don't be timid.

Embrace a little discomfort.

It's said fear and ignorance are good business, and that might be true. But good business for who? We don't have to let fear affect us. In fact the right kind of fear can be a good sign. It means we're on the right track. We're stepping outside that comfort zone. And ignorance, ignorance doesn't need to exist at all. It's almost as if ignorance is a choice. Why choose not to be informed? Why choose not to know?

If banishing fear and ignorance puts someone out of business, we probably didn't want them in business anyway.

As our level of convenience goes up, we tend to become more risk-averse. That's not the best direction. Keep the spark alive. Continue to push, to live, to take calculated chances. In this we're encouraging our future explorers, creators, inventors, to push the proverbial envelope.

The question becomes, how do we, the average space fans, contribute to that?

The best way for most of us is by being informed. Informing others. By getting out there and being part, spreading the word and, in so doing, sharing what we know.

There are those who will live their lives sitting at home, enjoying the fruits of the results of Those Who Dared.

Nothing wrong with that. These are quite likely the backbone of our society. Indeed, your humble authors fall mostly into that camp. We're certainly not out building rocket engines.

But we're *living*.

You should be too.

Optimism

WILLINGNESS TO TAKE THOSE RISKS REQUIRES OPTIMISM. A belief that you'll achieve the results you're after. We humans are great at this. We imagine great promise, great possibility, we act and ...

If we fail, if the results aren't what we expect, we revive hope and keep after it.

Just ask anyone who's ever bought a lottery ticket.

Eventually we *do* win. If not us directly, *someone*. Someone among us makes it. Typically that victory comes from multiple learning experiences, an equal number of failures, refinement of techniques, development of new methods and so on, leading to ... Success.

A quote on optimism from Kurzgesagt (Kurzgesagt means "said shortly" in German, or "in a nutshell"), one of the better channels on YouTube for those of us regular people curious about science stuff, sums it up nicely:

"Being optimistic about the future of humanity is not mainstream, and we think this is horrible. Pessimism often sounds smart, and gets more views, while optimism can sound naïve. But this is a bias that's not helpful for us as a species."

The Stakes

PUT ANOTHER WAY, we might ask that you "look for the good and praise it". Silence the negative chatter when you hear it, and certainly don't contribute to that buzz-kill.

This is an exciting time.

It's hard enough pushing the envelope of the technologies that will get us there, the things that will make our lives better, without people slinging poop at those who are trying to make those very things real. If we're not going to directly contribute by being one of Those Who Dare, we should at least applaud and celebrate those who are.

That fruit we're enjoying? That fruit we plan to enjoy?

Both depend on it.

And so don't sell yourself short. Just because you're not One Who Dares, you can dare in other ways. It's amazing what simply that, being brave enough to stand up and be a positive supporter, can do. Ambivalence is better than negativity, sure, so if you can't be helpful you could at least not be harmful. You could simply stay quiet. What's that cliché? If you don't have anything nice to say, don't say anything at all?

But what about *positive* support? Encouragement?

What about cultivating your own, positive outlook?

That's far more powerful than you think.

The News

EVEN THINGS LIKE HEADLINES can change the tone of those around us. Consider this example:

"Looks Like Another Failure. Launch Company Blows Up Another Rocket. More Money Wasted."

Versus this:

"Launch Company Marks Another Milestone. Will Use Results From Latest Effort To Add Features To New Model."

Pretty different vibe, eh?

On a more personal level, when you greet people, are you the one that opens with, "Did you hear about that terrible thing in Germany? Those pour tourists that got mauled by that circus tiger. Such a tragedy."

Or are you the one that opens with, "Did you see the new OSIRIS probe brought back an asteroid sample? Amazing. They're saying it will give us clues for tracking Near Earth Objects."

We say work on being that second person.

You may not be able to control the headlines, but you can control the things you say about them. You can enlighten your friends as to the real wins. Focus on the positive, minimize or outright ignore the negative. This is the way you encourage success.

There are a hundred sayings and examples of why this is.

Turn off the news if you can manage it. The big news networks aren't focused on anything positive anyway. Plus, less time spent there means more time you can spend with your favorite tech and space news sites. Or whatever fun thing you prefer.

In the mainstream news the science stories will be buried here and there, if at all, but on the tech sites they'll be headlines. The idea is to drive attention away from the negative and put that attention on the things we want to see more of.

Sadly the days of objective news reporting are gone. Even the weather has a spin for drama.

—

A short rant.
Kind of saw this coming, didn't you?

Dear people that actually have a platform and a responsibility:

If one's goal (presumably) is to generate interest in your chosen field (space things), to inspire a public that may not yet know much about the technical stuff you're covering, are you likely to get their support if you focus on the troubles that science is having?

If your audience already doesn't understand it clearly you'll only generate more opponents.

And what's the point in that?

Any source covering space and new technologies, advancements that are so necessary for our collective future but that are already poorly understood, should focus on the things that are being accomplished. On the gains that are being made as, yes, challenges and difficulties are faced and overcome.

Otherwise that source sucks.

Sorry to be blunt.

It's easy to think, Oh! But we need to know the bad things! That would probably be the retort of said sources.

"The people need to know how we're failing!"

No, not really.

Only way we need that information is when it's framed with solutions and ways we'll make it work. The *real* news is the news of the things that are being done to make it go right, the things we can actually do to help.

Everything has problems. Especially new, groundbreaking stuff. Everything. All of it.

Trouble, problems, difficulties.

We care only that we're solving those challenges, what we're doing to get there, and how we're reaching success.

Not what's going wrong.

Let's start seeing more headlines of *this* tone and timbre:

How we're making things go right.

It Ain't As Grim As All That

SO ... HOW *DO* YOU STAY INFORMED?

That's the trick, isn't it.

If we're going to ignore the negative, but negative spins are the majority way things are ever and have ever been presented ...

What's a poor, budding space enthusiast to do?

See the next section, *For Your Consideration.*

Obviously we need to at least be obliquely aware of things other than science and space. If you *totally* avoid knowing what's going on in the world you end up with your head in the sand. What we're advocating is keeping your head high, surveying the land without diving into the gritty minutia. As presented by the media there are tragedies and bad things *everywhere.*

Again, we invite you to look around.

Go outside.

Shake a few hands.

Touch grass.

The falseness of that grim outlook is hard to miss.

Mostly we're doing pretty damn fine, with a strong forecast for doing better.

For Your Consideration

ONE WAY OF STAYING INFORMED might be to find a source you prefer and use that. Skim other headlines in order to stay objective, but use that reliable source. Headlines will also generally inform you of what *is* important. If a bunch of headlines are talking about an invasion somewhere, then that's worth a deeper dive.

If it's just one alarming headline about a terrible raccoon attack somewhere in the world, unless you plan to do something about it, that one's probably worth avoiding.

Which brings us to a sort of litmus test for *any* news:

Don't read anything negative unless you plan to do something about it.

Let's be honest. You can do something if you choose to. No matter how extreme. You could get on a plane. You could send money. You could volunteer. Either choose to do something about the situation and do it, or choose not to and don't. All we're saying is don't read the negative news only so you can sit and think about how bad it all is.

Sounds extreme, and maybe it is, but if the goal is to take a more proactive role in our world (and it should be), we don't need to load our attention (and time) with things we can't, or don't intend to, do anything about.

No need even reading the details.

A set of hardcore "either/or" options for evaluating headlines might be:

1. Decide you can help. Read it and do something. Or,
2. Decide there's nothing you can or will do, ignore it and move on.

It's tough, we know, but if you use that criteria and start skipping such stories your thinking will clear, and you'll become that much more effective at solving the things you *choose* to solve.

Also, reading is better than watching. Committing yourself to any sort of news recording or, worse, broadcast, means you have to consume the story at the pace of the presenter, along with being subjected to the emotional spin with which they choose to dole it out. That spin often includes syrupy fear-mongering, veiled antagonism, calculated indignation (see the *IIS* section below) and worse.

To combat that we coined a phrase:

Reading.
It's the new watching.

With reading you set the pace. When reading you can read in the tone you choose, more easily skim and skip annoying parts, even pass over the dramatic sensationalism, digging out the real news.

Better, faster, easier, saner.

Stick to headlines. Skim for messaging. Read only the news you want. Avoid recordings and especially broadcasts.

Get on quickly with your day.

—

Way back when, when Facebook was a thing, as an experiment one of us tried to clean our feed by clicking Like only on benign things, like weddings and babies. It worked; after a while the FB algorithms adjusted and were showing pretty much nothing but ads, babies and weddings. It ended up being weird, and mostly useless, but the principle is sound. The cyber systems running things aren't nefarious. They're simply programmed to present the things you show interest in—expecting that you'll keep scrolling if you see more of them, which means they can show you more ads.

Okay. So maybe a little nefarious.

Morale

THE CUSTODIANS OF THE COMMUNICATION LINES of the world, the sources of our information and news, get to decide what we're fed, what we're made aware of—basically, they get to shape the narratives that influence our lives. What we hear we hear because they decide we should hear it. What we don't hear, well, they decide that too.

The idea of simply reporting the facts, whatever those facts may be, and letting us form our own opinions is long gone. If it was ever really a thing.

Back in WWII those custodians of the news were steered by the British government toward not making headlines of disheartening stories, focusing instead on reporting events more likely to boost morale. This tactic was used to good effect. It kept the minds of the populace on victory, not setbacks. We have to believe that positive outlook influenced the overall outcome of the war.

It's still manipulation, but at least it was manipulation for the greater good. Too much relentless reporting of terrible things becomes a form of propaganda in its own right, convincing people "it's all bad" and that there's no hope.

What's the point in that?

(The point, of course, is multi-layered and serves the purpose of keeping us in our lanes, but we've already gone into that.)

We ask that question only to shine a light on the destructiveness such an SOP (Standard Operating Procedure) has on our enthusiasm for life. The ones pushing those narratives would seem to prefer us suppressed and controlled; again, for reasons of their own.

The majority of us definitely do *not* want that, nor do we gain from it.

Terrible things do happen, yes. Grim things. Bleak things. We need to know about them only to the extent that we can address them. Terrible things must be confronted and handled, yes.

But they don't need to be harped on beyond their solution.

The preponderance of our focus can't be on the bad. (And yes, we did just use preponderance in a sentence.)

Our morale depends on keeping our eyes on the prize.

Our morale depends on our positive action, and the news of the positive actions of others.

There's plenty of that to be reported on.

We dare say there's way more of that source material to draw from than the other.

IIS

INSTANT INDIGNATION SYNDROME. Let's call it "IIS" (not to be confused with the ISS, the International Space Station), the condition whereby we react, often without thinking, to any bit of alarming or contentious news that goes against our decided-on beliefs, or which reinforces and aligns with our fixed ideas. It affects many of us, and as individuals we aren't wholly to blame.

IIS has come about through a relentless news cycle that pits us against each other's ideals. It's a knee-jerk reaction, and it's been beautifully instilled in all of us. More often than not those ideals have been fabricated for us, and we're usually given a This-or-That option, forcing us to side with one indignant group or the other, programmed nicely to spark Instant Indignation when the right buttons are pushed.

Usually we don't even really stop to think about what we're getting indignant about.

Again, not wholly our fault. It's how the system works, it's how it's been designed, and it's important we behave that way. At least it's important for the current system to continue. Our strings have been tied (we've allowed them to be tied), and if we stop responding when they're pulled the whole thing starts to get shaky.

Ultimately that would be a good thing.

A disruption of the current easy-divisiveness, the instant indignation, would begin to shift us toward considered thought and honest, unbiased evaluation of information. Open discussion, mindful consideration of facts—basically, all the things that would collapse Media and Politics as we know it.

Think we could suffer through that?

Wouldn't you agree it's time for objective consideration of the issues?

Offense comes easily these days, it seems.

Let's change that.

Conflict Sells

YOUR AUTHORS HAVE DABBLED in screenwriting, video games, other fictional dialogues, and at least one of us has written a few sci-fi books, so we know the value of conflict.

Conflict sells.

Remember that skit from *Family Guy*? It was called "The Even Couple", a play on the old TV show 'The Odd Couple". In it we see quite clearly, and quite hilariously, how important conflict is for entertainment.

Conflict sells.

Controversy.

So yeah, we get it. Not only do we get it, we've actively cranked it out in the form of screenplays, books and games by the bucket loads. Conflict is juicy.

Conflict keeps us glued to the story.

Conflict is critical for fiction.

Storytelling is nothing without it.

Without it we put the book down. We turn off the screen. We walk away. Can you imagine a video game with no challenges? No boss fight? Conflict is what draws us like moths to the flame.

Which is exactly why the news exploits it.

It's why we watch it.

Which is, therefore, why the news keeps doing it.

But come on.

Criminy.

Can we save the drama for our collective mamas? Can we keep the conflict in our fiction, where it belongs? Can we, instead, focus on the positive in our actual, real lives?

Let's start paying attention to the news where we're winning.

There's plenty of that happening all around the world.

DID YOU KNOW?

In 2021 *Inspiration4* marked the world's first all-civilian mission to orbit.

This was *huge* news.

Or at least it should've been.

Named in recognition of the four-person crew that raised awareness—and over $240 million—for St. Jude Children's Research Hospital, this remarkable mission flew to orbit with a crew of civilians, only civilians, who then stayed in orbit for three days.

One of them was a physicians assistant from the hospital itself, Hayley Arceneaux. Hayley, who herself suffered childhood cancer and was saved by the people at the very hospital where she now works, not only became an astronaut, she was instrumental in helping to raise the quarter of a *billion* dollars that went toward the St. Jude mission.

Civilians trained to become astronauts. They flew to space alone. They stayed there three days. They came back safely, on their own. They made a charity out of it to boot, helping one of the most worthy causes on the planet.

Yeah.

Epic. Groundbreaking. Monumental. History-making.

All the adjectives and hyphenated adjectives.

Netflix even did a special on it.

Where were the headlines?

Hm.

And this is our point. As consumers of the news these are the things we need to insist on seeing.

Human Nature

THE TRAIN-WRECK ALLURE that draws us to negative headlines seems built-in and unlikely to go away. We crave that conflict, that drama, the juicy gossip about the neighbors.

It's very human to be fascinated by tragedy.

There's an inborn predisposition for this, driven by nature. The animal kingdom does it. It's called predator inspection. The act of getting close to danger, like a gazelle hedging closer to a cheetah; observing the frightening in order to get a gauge of the threat, and an idea of how it might be handled. Think of the last time you couldn't resist peering over the edge of a tall building.

When it comes to us humans it once served a greater purpose. Now we need it less. Now we have the means to educate ourselves as to various dangers and formulate responses, without the need for ogling. Now we're evolved enough to fight the urge, able to consciously resist the impulse to look when we know we really have no business looking. To drive by the flashing lights at the accident scene and keep our eyes on the road.

The police, the fire department, the ambulances, they're all there. You're not needed. Keep moving.

Yet ... that impulse is still deeply ingrained within us.

The news media relies on it.

One snapshot of the swath of destruction a tornado can inflict is enough. Maybe not even that. Pair that with some helpful information on what to do, how to be prepared, etc., and that's as far as the value of inspecting the danger of tornadoes goes.

Can you do something to help?

Will you?

If not, move along.

Those are people's ruined lives you're getting a thrill watching.

At this stage of our evolution we have the means to become intelligently informed and make rational, not emotional, decisions. We don't need a relentless slideshow of the juiciest, cherry-picked aftermath photos to learn more about how destructive a twister can be. We get the message.

Still, we can't help but look.

Though the concept has survival value, the way this natural human impulse is used to manipulate us only stokes fear and raises blood pressure. The opportunities for doomscrolling we're fed 'round the clock fill that urge beautifully. But it's empty calories. We gain nothing from such morbid fascination—beyond what information can be gleaned about safeguarding ourselves against similar catastrophe.

And that doesn't take much.

The rest is gluttony.

We crave it, the same way we crave another candy bar. Yet, like a candy bar, it's equally pointless to our existence. The fact that we even have a term like "doomscrolling" in our modern lexicon is a bad enough sign. We don't have to perpetuate it.

(Okay, perhaps arguments could be made for the point of candy bars. A world without candy bars would be a sad world indeed.)

Again, if you can help, in any capacity, please do. Be useful. Otherwise don't ogle.

Tragedy does not need spectators.

Sports, air shows, plays, concerts, the opera—these are the places to be a spectator.

Not life.

Life needs participants.

Get involved, help fix things, or stop staring. Work on something else.

What If ... ?

THERE'S A SAYING: "Coups and earthquakes sell newspapers."

But it doesn't need to be that way. We could take control. In the interest of putting those macabre desires aside (at least when it comes to reality; no one's saying we should stop watching horror movies, tragic dramas, kick-ass action flicks—you get the idea), we could choose instead to focus on the positive. What would happen if we only paid attention to the things that have us winning? Could we shift those negative headlines to page 3, 4 or 12, moving the *good* news, to page 1 and 2?

Instead of having to scroll to find the good stuff, let us instead scroll long and far to get the negative news. The bad stuff. The drama and the trauma we can't seem to get enough of.

Can you imagine?

You go to your favorite news source and on the front page are amazing new discoveries, achievements (like the *Inspiration4* story), sports wins, stories of human kindness, successes in space and technology, winning business news, headway being made in fields of human advancement and so on.

Where's the bad stuff?

Of course it's still there. The bad news still needs to be reported. But in the same way the good news also has to be reported, we flip the script. The good news is no longer buried beneath the rest, the bad news is. The good news is front and center. Where's the bad news? You have to work your way down to find it. Now *it's* the stuff buried. It's not at the top.

The good news is.

Now you're "winscrolling", seeing nothing but wins. Successes. Positive achievements. Good news.

That's a suggestion that will probably never see the light of day, but how cool would that be?

Instead of having to hunt for the tech news, the news of progress, discovery and advancement, it would be right there, at the top of our news feeds.

Make us hunt for the other stuff.

Who knows?

After a while maybe we'd get tired of looking for it.

Our Native State

AS A TEST OF THIS PRINCIPLE, next time you're talking with someone who seems mostly negative about things, ask yourself:

1. How much of that is their own, true outlook?
2. How much of that is, or has been, influenced by the negative energy being fed them by the media?

We're all subject to that steady bombardment.

As we get older, it only piles up.

Illustrating this contrast, have you noticed few children are anything less than wildly optimistic? Enthusiasm, energy, optimism, positive outlook—these characteristics define what you might call our native state.

We're born with those traits.

Good news is, we never truly lose them. It's only over time that we begin to go into agreement with the negative energy thrown our way. The positive drive is always there.

The positive *is* us.

It only gets beaten down.

Battered, until we have a hard time reviving it.

Yet ... it never leaves. It *can't* leave, because it's who we are.

The rest is just crap that's been piled on.

Strip away that programming and ...

Voila.

We're ourselves again.

Our space future depends on us being able to dust off that child-like wonder and enthusiasm, restoring the core of our character once more.

—

Curb Your Enthusiasm, a long-running show on TV—as a title—is funny because that general sentiment is pervasive throughout society. The show itself is funny too, for similar reasons.

Curb your enthusiasm.

Settle down.

Rein it in.

Easy there.

Wanting others—or worse, ourselves—to be less enthusiastic is not a native sentiment, yet it exists quite commonly. A pervasive tamping down of our eagerness for life, both external and self-inflicted. We're taught failure is an end, when in fact failure quite often marks an opportunity to learn and improve, to make a fresh, better-informed beginning, and so we doubt ourselves or guard against our own excitement. Others question our abilities, express doubt or otherwise sneakily layer on subtle negative inputs. Inputs intended, whether subconsciously or consciously, to dim the light of our natural enthusiasm, to discourage our curiosity.

To slow us down.

Keep us in our lanes.

Encourage us not to reach so big.

Hold expectations to the level of "reasonable".

Don't work too hard!

Probably one of the best, and most common, ways to "ease back the throttle" of all time.

How many times have you heard someone say that?

Don't work too hard!

How many times have you said it?

Funny if said in jest (which it usually is), not so much if said with genuine intent. Pause a moment to deconstruct that one.

Working hard is *exactly* what we want to be doing. No one says it can't be the most fun we've ever had. Done right work *should* be fun. We get energy from our motivation, passion from our purpose. Working, production, accomplishment ... these are the very core of happiness.

R&R is vital, of course. Cool those heels. Enjoy that tasty beverage and the company of friends; downtime, filled with the satisfaction of a job well done, tinged with the anticipation of the next great adventure you're about to throw yourself into. The next great result you plan to achieve.

The rest of us are counting on you.

Your contributions mean more than you might realize.

The irony is, those who say these things—who may even believe they mean well—are subject to the same, subtle suppression. They're going through it just like we are, living, of course, in the same society where it prevails. Many times they're just echoing things they themselves have been told.

Where it starts, no one knows. But we don't have to pile on. We can refuse to "curb our enthusiasm".

We may not have started it, but we can be the ones to bring it to an end.

When it comes to space this is especially important.

Never be afraid to shine.

Culture

CULTURE PLAYS A ROLE IN THIS. The values we instill. It's up to us to create a culture of enthusiasm and cooperation—not only for ourselves, but for the generations in our charge. The next batch of humans we introduce to the world will depend on us to set the stage for the future they take over.

It's up to us.

Whether you agree with the following statement or not, the reality is, whatever's going on in the world—if it has to do with humanity—we either:

- Let it happen, or
- We made it happen.

This goes for the bad *and* the good.
Everything good, everything amazing, every great accomplishment, every great new member of the next generation that was raised right and shown a world of hope ...
All of it us.
This world is ours and only ours.
Think of your house. What happens in your house is the responsibility of everyone living there. No one else.
Earth is our house.
What happens here is our responsibility.
A responsibility that scales.
See to yourself, your area, your people.
The friction in the world can seem overwhelming, but it isn't if we each do our part. It isn't if we each decide to take responsibility. It isn't if we each make the decision to build the positive into our respective cultures, to nurture and instill strong values of community and purpose, hope and inspiration in ourselves and, vitally, in the up-and-coming generations in our charge.
Forty Suns is about taking control of our destiny. About no longer agreeing to be victims. Perhaps more importantly, it's about no longer *wanting* to be victims.
Victims, after all, must be *given* solutions.
They don't *create* solutions.
We'll never make it from a victim point-of-view.
Forty Suns is about change.
One by one, starting with each of us. If you make it a habit of refusing to behave badly you start a trend. If you make it a habit of

taking responsibility, others take note. If you make it a habit of refusing to hear ill, those negative people around you will either stop talking to you (or you them), or they'll snap out of it and realize the choice to be negative is their own, and they're free to change their mind.

Each convert one more shoulder to the wheel.

One by one.

Perspective

PERSPECTIVE HELPS.

Battles over resources, territory and power are normal and to be expected. Kids at the playground do it. So do us grownups. We're not suggesting we change human nature, or even that we'd want to. Less fighting might be better, sure, but at the end of the day a little heat, a little friction, is what keeps us honest. Disagreements drive the passions that craft our world. Competition keeps us striving for better things.

Put another way, beat too many swords into ploughshares (the cutting blade of a plow), and we might one day find ourselves wishing we still had a few swords.

The ability to fight is important.

Yet we can't forget:

Until we solve the one-world dilemma (the whole point of *Forty Suns*), this planet is the only one we have.

Which makes things a bit different.

At least for a while.

We'll have more opportunities to fight in the future. As we already mentioned we could keep warring amongst ourselves, even declare an ultimate winner, but right now, with the means we have at our collective disposal, it seems it would be way better to focus on finding alternatives to this sandbox, to work on discovering new playgrounds. A little more friendly competition, until we can establish ourselves on a few new worlds.

That will take time, no matter who's in control. Perhaps it's foolish to think we'll stop fighting long enough to become truly extraplanetary. Still, it's worth keeping in the forefront of our mind.

A little less focus on the squabbles, a little more focus on cooperation, may be all we need. A little more working together, a little less fighting.

We could put it this way:

- 100% cooperation is unlikely. In fact we probably don't want everyone agreeing to everything.
- 0% fighting is also unlikely. We don't want that either.

But let's be better.

A quote from former Space Marine and current *Forty Suns* spokesperson, Tai Rade, sums it up:

"There will be plenty of time in our future for space wars, cold wars, star wars, race wars, culture wars, cola wars and other wars, trust me, but maybe, for now, at least for a little while, it's time to put the slap-fighting on hold and be friendly. Focus on setting up that next playing field. Otherwise, if we lose the only one we have, it's pretty much game over."

Our Brotherhood

THERE'S A LOT TO ARGUE ABOUT. There always will be. There's also a lot to agree on. That, too, will always be the case. With anything there will also always be that small minority that stand to lose if the rest of us are united toward a common goal. Division, suspicion, fear—as mentioned earlier, these things do benefit a select few, and so there will always be covert resistance to improvement.

It's up to us to push past those distractions and focus on the things that matter.

Remember most things are done out of benefit for the entity doing them. People, individuals, can be altruistic, thinking of others as well as themselves. Governments, businesses and other entities don't tend toward that.

So what's the benefit of keeping us divided?

Would there be a benefit to making us think we *can't* achieve an extraplanetary existence?

Surely it would be in the best interest of some that we not think that way. Surely some would see value in attacking any effort to help humanity advance.

Good news is there's a way to fight such bad intent. The way to combat those people is relatively simple.

Win.

Succeed.

Flourish and prosper.

In fact we'll go one step further and say you'll know for sure you're doing the right thing if you're getting attacked. Success, by its very nature, comes with attacks. To be successful you need to be willing to stand up and be hit.

Because the haters *will* hit.

It's what they do.

Rather than resist or hope against those attacks, welcome that milestone. Thank the haters. Because by targeting you they've confirmed one simple, yet monumental, truth:

You made it.

You're a success.

And don't worry. The rest of us will celebrate your wins. The sane among us (which, believe it or not, is most of us—see *The 90%* below) realize we're in this together.

Look around, at your compatriots, your fellow humans on this planet. Both near and far. Their gains are your gains. When they win you win—and we mean that in a positive way. When others are doing positive things in the world, when others are achieving positive gains, it's a win for all of us. Which also means when you win the rest of us do as well.

Therefore celebrate the gains of others.

And make your own successes known.

We are, after all, in a brotherhood with each other.

We owe it to each other, and ourselves, to get—and keep—the show on the road.

Like a fire truck being chased by a dog, if we're too busy making progress, clanging the bell, directing enormous energy toward solving problem after problem, roaring down the street as we head for victory, those that would drag us down tend to become meaningless—if we even notice them at all—as they fall further and further behind.

Taking Ownership

ON THE SUBJECT OF THAT BROTHERHOOD, when multiple parties are involved responsibility for a thing is rarely 50-50. Nor is it ever 100-0. Each party shares their portion of accountability. It may only be 99-1, but each party's hand is there.

No one is every fully absent responsibility.

No matter how righteous you might feel—and you might be quite justified in that feeling—you share your responsibility in any outcome that affects you. (Perhaps even some outcomes that don't; a mind-blowing idea.)

No one and no thing is ever 0% at fault.

It's important to recognize—and own up to—that 1%, if that's all the responsibility you share. That 1% is still your burden to bear. Or 5%, or 60%, or whatever portion is yours. That's true for the bad *and* the good. Own (and learn from) your mistakes, celebrate (and learn from) your wins. Luck and misfortune are not unreal, and we have a hand in each. Who can say? Maybe you just decided to be lucky. Or unlucky.

Take ownership.

We can apply this to our space future. You may not be a rocket scientist, taking high levels of responsibility for that future, but everything we do matters. Being a fan, spreading the word, getting people talking about the things that will get us there ... that, in itself, is taking responsibility.

If you're doing that, we salute you.

Understanding—and taking—ownership is the true path to freedom, and to greatness.

The 90%

IF YOU PUT A HUNDRED OF US IN A ROOM, with no outside influence, at the end of the day 90 of us (or more) would mostly agree on all the important stuff. The things that really matter. We're not suggesting those 90 are without fault, or that we don't each have our quirks, our idiosyncrasies, the things that sometimes drive others up the wall or make them scratch their head and ask, "Why? Why are you like that?"

But those things are what make us individuals.

Those 90 (us), in their hearts, are good.

We are good.

Those 90 are the ones we can, and do, and will, work with.

Yet, through the bias of media (social and otherwise), the voices of the other 10 somehow end up the loudest, and our perception slowly becomes that the world is split down the middle.

In that world-view half of us think one extreme way, half of us the other.

Can we truly be that divided?

That's a negative, Ghost Rider.

Let us once again suggest you look around as you go through your day. Observe the people in your world. Not the people behind screens; the people walking around in 3D, the ones you see live and in color. At the store, at the mall, at the club. Doesn't seem that way in real life, does it? For at least 90% of you the world doesn't look anywhere near as divided as in the news. Certainly not as hopeless.

We hope-having people are in the majority.

People may act bad; it doesn't mean they *are* bad.

Circumstances, upbringing, life—many factors go into a person's choices. Most of us started out with, if not noble, at least positive intent. Search your feelings, you wonderful person. You know it to be true.

Even Vader found a little redemption in the end.

Strip away the crap and, fundamentally, the majority of us are good people. Which means there's plenty of room for lofty aspirations. Quite a bit, actually.

It's therefore up to us, those of us who know better, to maintain a positive focus, while bringing that positivity to our actions. Media will continue digging for (and finding) that 10% of misery, then make shoutingly bad-scary negative headlines out of it.

Don't buy it.

Again, if you can do something about it, do it.

Otherwise don't get drawn into the negative spiral. Let's make it an *upward* spiral. Introduce a touch of what's been charmingly called "toxic positivity".

Your positive actions *will* have an impact.

Contrary to the imagery we're presented daily, there is no All This or All That. Sanity—and most of the world—exists somewhere in the middle. Sanity exists with the 90%.

We, the sane, the able, the ones taking positive action—at every strata of existence—are the builders of the future.

On What Could We Agree?

WE AGREE THERE'S DISSENTION IN THE WORLD. Yay! That's one thing.

What are a few positive things we could agree on?

Here are a few ideas to which the sane among us (that 90%) could likely agree.

- Help is possible.
- Helping people is good.
- Getting help is also good.
- Stopping others from hurting others is good.
- The wrong thing to do is nothing.
- Protecting this planet is good.
- Understanding more of the world around us = good.
- Expanding our reach to other worlds = good.

We might also agree to a sort of umbrella statement, like this one from Marcus Aurelius (one of the good Roman Emperors):

"If it is not right do not do it; if it is not true do not say it."

A solid concept on which most sane people would agree.

That's a handful to start. We'd love to hear any ideas you have, perhaps for inclusion in a future *Edition*—definitely to post on the site.

A Hypothetical Round Table

WHEN IT COMES TO FINDING things on which to agree, how cool, and potentially strange, would it be to be sitting at a discussion table with that one person you thought you'd never agree with, thought you could never speak to or listen to, so different, so unlike you, and yet there you are, talking with them, having an actual conversation about some next great space enterprise?

It could happen.

This, ultimately, might be the greatest agreement we could all make. To agree to talk. To consider the facts honestly, the possibilities. To openly discuss ways to make our collective space future a reality.

Could we have that conversation?

The Value Of Talking

IT'S BEEN SAID COMMUNICATION is the universal solvent. Like water it wears away barriers and smoothes flows, eventually creating a vital conduit for the easy exchange of ideas. As such, communication may be the greatest tool we have. For any success, in any area.

What say we use it?

Not only more wisely, but actually *use* it.

We can't yell forever, though some would try. If two people just stood there and kept yelling at each other (and didn't walk away while they were still yelling), eventually the yelling would turn to angry talking and, if they kept at it (and didn't walk away while they were still angry-talking), eventually some sort of rough discourse would evolve and, if they kept at *that*, eventually, actual ideas would be exchanged and, eventually eventually, agreements would spring up.

Maybe even a new friendship or alliance.

A remarkable truth.

Freedoms

FREEDOM IS GREAT. That's probably something else we could all agree on, right? More freedoms = mo-better-good. In fact, being free from everything would be the ultimate freedom. That should be something we ...

Wait.

Hm.

What say we pause for another philosophical moment?

Freedom is great, that's quite true, but barriers and rules serve us well.

Balance.

Seeking the perfect balance is the eternal struggle.

How to have ultimate freedom, while at the same time having a game to play, is the most central quest in our human existence.

In yet more of our sports comparisons, take (once again) football. How kooky would it be without sidelines, end zones, yard markers and a way to keep score? What the heck would it even be?

And so, while we should always and forever be pushing for more freedoms, while also defending the ones we have (eternal vigilance is the price of freedom, after all), we need well-crafted guides within which to play.

The right balance will set us free.

We do get it right from time to time. Sometimes we even nail it. A perfect balance of challenge and motivation, a perfect balance of difficulty and success, struggle and reward, chocolate and peanut butter—you name it, we do our best when we strike that harmonious sweet-spot. When we name our playing field, name our rules, then cut loose with everything we have.

That's when we all win.

Honestly, that's when we have the most fun.

Much has been done already to create the world we live in. Let's figure out how to keep making it better. Don't let go of freedoms. As we've seen time and again, once you give up a freedom you play hell getting it again. If you even can. Freedoms are tough to claw back later when you want them, which makes giving them up its own kind of dwindling spiral.

Don't do it.

Instead let's figure out how to make the best use of the freedoms we have, while creating trust and responsibility among ourselves.

Doing some of the things we talk about in this book will lead there.

Let's make this a bigger, better game.
For all of us.

Soul Of The Frontier

THERE ARE ONLY TWO PATHS FOR ANY ENTITY TO TAKE, whether an organic life form or otherwise:
Expansion or Decline.
One direction has an end. The other is potentially limitless.
Those are our options.
There are probably many universal maxims, but one of them for sure is:

Nothing Stands Still

There is no steady-state. Not one that lasts.
Expansion is the theme of this book. Specifically expansion into the next frontier, space. Humanity needs an escape hatch, and not just a place for the current crop of us to flee if we have to in an emergency. Together we need to be creating new homes and new possibilities.
Our history is filled with examples of how well we do this.
One recent period, the can-do spirit and camaraderie of the old American West, is a case in point.
Imagine the once-reality of the following quote (yes, more quotes) from American poet, Arthur Chapman:

"Out where the handclasp's a little stronger, out where the smile dwells a little longer, that's where the West begins."

Substitute 'the West' with 'space', and you get a sense of the shared frontier-spirit that can grip us.
We humans like discovery, and we like to do it together. Mister Chapman reminds us that, as we move into those vast unknowns, filled with hope, promise, it's reassuring to know there are others with us who have our back.

—

When it comes to acting on that spirit, it's been pointed out that we're the middle children of history. Born too late to explore the earth, born too early to explore the universe.

A clear dividing line exists.

And we've reached it.

In the same way the earth was once a wide and unknown canvas to the people who first walked out of Africa, filled with the possibility of expanding everything they knew, so too is space a wide-open frontier we're only just beginning to understand.

Almost as if, at this moment in history, it's less that we're the middle children, more that we're the crew manning the fulcrum of humanity's future. We're the ones resetting our eternal mandate to explore.

We, those of us alive today, sit at the cusp of a new era.

Whether we explore the universe in our lifetime, future generations will look to us as the ones who took the first steps to make that possible. We will be the ones to cross that line, with conviction, throwing open the way to the stars.

Together

YOU DON'T CHANGE MANY MINDS ARGUING. If any. Trust us on that. Nor can you mandate true cooperation. Epiphanies are personal, and each of us will or won't experience our own. Each individual must come to their own conclusion, their own decision, before taking effective action. Yet, even though no one can mandate cooperation, cooperation is the only way we get there.

It starts with each of us.

True understanding is a personal thing. Not intellectual awareness but true, visceral understanding of what's right and what's wrong. Right leads toward survival.

You're unlikely to wake anyone up (here we mean make them awake to their environment, not make them "woke"), by:

- Lecturing them.
- Boring them.

- Arguing with them.
- Failing to take into account their current point of view.
- Shouting.
- Disguising a desired ideology.
- Preaching.
- Refusing to listen.

(Hopefully we haven't done too much of any of those, though we know we probably have.)

Let's wake ourselves up to the tremendous possibility ahead.

Again, people *are* good and *want* good.

Our aim should be to wake ourselves to the fact that Positive Gain can become our new, shared reality.

Failure need not be the default.

To add to that, let us point out there's truth to that old adage:

No one ever really fails. They just quit.

And so, as we agree to cooperate, let's also agree not to quit. Shared experience is one of our most important tools, so let's get together and share space. Space Age 2.0 is here, and we're part of it. Each of us. Let's see it through this time.

Together.

We made it. We arrived. Somehow, some way, we managed to get here. However we did it, however those before us did it, whoever stepped on whoever's toes, whoever helped whoever, whoever hurt whoever, whoever is here as a result of the work of someone else and not their own, whoever has suffered as a result of the actions of someone else, whoever conquered whoever or whoever was conquered by whoever else ...

Dang that's a lot of whoevers.

Point is, we're here.

All of us.

Together.

Take a look around. We keep saying that, but we really want you to *look*. This is the crew. Team Earth. Love them or hate them, these are your compatriots. We hope you'll start loving them. These generations, this and the next, are the ones that will get it done. What happens now happens on our watch.

Past wrongs can be fixed, if need be, surely, but time spent dwelling on such things is time lost getting the show on the road. The past is behind us. Better to make our future. As well, spending too much time dreaming about the *far* future, while valuable, can also get in the way of getting things done *now*.

Now more than ever we need steady progress.

Maybe a handshake. Maybe a fresh acknowledgement; a little amnesty, human to human (we're all human, after all); forgiveness for whatevers, with an understanding that better is to come. Followed by a fresh agreement to look ahead, not behind. To What Can Be, not What Was.

It's only as complicated as we choose to make it.

We've all been bad to each other. That much is true. We've also been good to each other. For most of us, most of the time, the latter is the greater truth. Let's remember that, then get shoulder to shoulder and push.

All of us.
Each of us.
Together.
Forward.

Why We're Here

WE'LL TELL YOU, BUT WE'RE GOING TO DO IT WITH QUOTES.

Did you really think we were done with them?

Forgive us, but smart people say smart things and, sometimes, their quotes are a great way to make a simple point.

Here's one from the band Shinedown. It's a bit sappy, but the meaning is hard to dismiss.

At the end of the day one thing matters:

"No one gets out alive, every day is do or die / The one thing you leave behind / Is how did you love?"

Which is to say: Love is at once a pinnacle, and a foundation. (Maybe someone will quote *us* on that one?)

Ozzy Osbourne (the Ozzman!) backs that up:

"Maybe it's not too late / To learn how to love / And forget how to hate."

Each of those, and hundreds like them, remind us to think on why we're here. If you really stop to ponder, you'll probably come to the same conclusion:
We're here for each other.
A world without the rest of us would be a sad world indeed.
Who would you talk to?
Where would you send holiday cards?
No matter that we might at times feel selfish, no matter any flashes of cynicism, in our best moments (which are the moments where we're truly ourselves) we realize this life is about more than just ourselves. More than just our own survival.
It's about everyone.
All of us.
The looming significance of our shared, one-world problem makes it clear that a "win" for all is a win for each.
Indulge your humble authors for another?
Related in its own way, this quote not from an astronaut, a political hero, a worthy scientist, wise rock band or any other, expected font of insight.
Rather, let us leave you with perhaps the wisest words ever spoken, by two of the most bodacious adventurers of all time.
Bill & Ted.

"Be excellent to each other."

Code Of A Future-Builder

IN LIFE THERE ARE WRITTEN CODES, unwritten codes, codes of honor, moral codes, codes of conduct ... whether officially outlined or not, we all operate by codes.

In the end a code, like an epiphany, is a personal thing.

Here are a few points of what we feel might comprise a code for the builders of the future:

1. Believe in our extraplanetary destiny.
2. Believe in your fellow humans (in particular the 90% that mean well).
3. Promote the successes of those engaged in building our space future.
4. Do your part to make that future.

Maybe we'll add to it over time. Your own thoughts are welcome. For now it works.

From Us To You

WHICH LEADS US TO THE END OF THIS WORDY TREATISE.
And an acknowledgement.

You're here.

We made note of that fact before, but let us say it again:

You're *here.*

For that we thank you.
Not just for being here, reading this book, or especially this long, and final, appendix (though indeed, thank you for that), but *here.*
On this world. With the rest of us.
Wherever you're at on your journey, wherever you are on this planet, you made it.
You're here.
Thank you.
For arriving. At this moment. For holding within you a potentially unlimited promise to bring life to our destiny in space. To bring change. Improvement. To make better the things that are good, to correct the things that aren't.
For being the hero of your story.

For adding that story to the rest.

In whatever role you take, what you've done already, simply *being* here, alive, filled with the ability to impact our collective future, charged with the power to help ...

For that we say simply:

Thank you.

Made in the USA
Columbia, SC
15 January 2025